SCHOOLING ACROSS THE GLOBE

Schooling matters. The authors' professional pursuits for over 25 years have been focused on measuring one key aspect of schooling: the curriculum – what students are expected to study and what they spend their time studying. This documents their conviction that schools and schooling play a vital and defining role in what students know and are able to do with respect to mathematics and science.

This research examines 17 international studies of mathematics and science to provide a nuanced comparative education study. Whilst including multiple measures of students' family and home backgrounds, these studies measure the substance of the curriculum students study which has been shown to have a strong relationship with student performance. Such studies have demonstrated the interrelatedness of student background and curriculum. Student background influences their opportunities to learn and their achievements, yet their schooling can have even greater significance.

WILLIAM H. SCHMIDT is a University Distinguished Professor at Michigan State University. He previously served as National Research Coordinator and Executive Director of the US TIMSS National Center.

RICHARD T. HOUANG is a Senior Researcher and the Director of Research for the Center of the Study of Curriculum Policy at Michigan State University.

LELAND S. COGAN is a Senior Researcher with the Center for the Study of Curriculum Policy at Michigan State University and was the U.S. Assistant Director for the Teacher Education Study in Mathematics (US TEDS-M).

MICHELLE L. SOLORIO is a PhD education policy student at Michigan State University.

EDUCATIONAL AND PSYCHOLOGICAL
TESTING IN A GLOBAL CONTEXT

Editor
Neal Schmitt, *Michigan State University*

The Educational and Psychological Testing in a Global Context series features advanced theory, research, and practice in the areas of international testing and assessment in psychology, education, counseling, organizational behavior, human resource management and all related disciplines. It aims to explore, in great depth, the national and cultural idiosyncrasies of test use and how they affect the psychometric quality of assessments and the decisions made on the basis of measures. Our hope is to contribute to the quality of measurement and to facilitate the work of professionals who must use practices or measures with which they may be unfamiliar or adapt familiar measures to a local context.

SCHOOLING ACROSS THE GLOBE

What We Have Learned from 60 Years of Mathematics and Science International Assessments

WILLIAM H. SCHMIDT
College of Education, Michigan State University

RICHARD T. HOUANG
College of Education, Michigan State University

LELAND S. COGAN
College of Education, Michigan State University

MICHELLE L. SOLORIO
College of Education, Michigan State University

CAMBRIDGE
UNIVERSITY PRESS

University Printing House, Cambridge CB2 8BS, United Kingdom

One Liberty Plaza, 20th Floor, New York, NY 10006, USA

477 Williamstown Road, Port Melbourne, VIC 3207, Australia

314–321, 3rd Floor, Plot 3, Splendor Forum, Jasola District Centre, New Delhi – 110025, India

79 Anson Road, #06–04/06, Singapore 079906

Cambridge University Press is part of the University of Cambridge.

It furthers the University's mission by disseminating knowledge in the pursuit of education, learning, and research at the highest international levels of excellence.

www.cambridge.org
Information on this title: www.cambridge.org/9781107170902
DOI: 10.1017/9781316758830

© William H. Schmidt, Richard T. Houang, Leland S. Cogan and Michelle L. Solorio 2019

This publication is in copyright. Subject to statutory exception and to the provisions of relevant collective licensing agreements, no reproduction of any part may take place without the written permission of Cambridge University Press.

First published 2019

Printed and bound in Great Britain by Clays Ltd, Elcograf S.p.A.

A catalogue record for this publication is available from the British Library.

Library of Congress Cataloging-in-Publication Data
NAMES: Schmidt, William H., author. | Houang, Richard Ting-Ku, author. | Cogan, Leland S. (Leland Scott), author. | Solorio, Michelle L., author.
TITLE: Schooling across the globe : what we have learned from 60 years of mathematics and science international assessments / William H. Schmidt, College of Education, Michigan State University, Richard T. Houang, College of Education, Michigan State University, Leland S. Cogan, College of Education, Michigan State University, Michelle L. Solorio, College of Education, Michigan State University.
DESCRIPTION: Cambridge, United Kingdom ; New York, NY, USA : Cambridge University Press, 2018. | Series: Educational and psychological testing in a global context | Includes bibliographical references and index.
IDENTIFIERS: LCCN 2018021301 | ISBN 9781107170902 (hardback : alk. paper) | ISBN 9781316621844 (pbk. : alk. paper)
SUBJECTS: LCSH: Mathematics–Study and teaching–Curricula. | Mathematics–Study and teaching–Cross-cultural studies. | Mathematics–Study and teaching–Longitudinal studies. | Science–Study and teaching–Curricula. | Science–Study and teaching–Cross-cultural studies. | Science–Study and teaching–Longitudinal studies.
CLASSIFICATION: LCC QA11.2 .S35 2018 | DDC 510.71–dc23
LC record available at https://lccn.loc.gov/2018021301

ISBN 978-1-107-17090-2 Hardback

Cambridge University Press has no responsibility for the persistence or accuracy of URLs for external or third-party internet websites referred to in this publication and does not guarantee that any content on such websites is, or will remain, accurate or appropriate.

David E. Wiley

We dedicate this book to my mentor at the University of Chicago who taught me what it means to thoughtfully and cleverly analyze data as opposed to just doing statistical analyses. He brought that mind-set to work on the Third International Mathematics and Science Study where my colleagues, Richard Houang and Leland Cogan, also came to appreciate his deep understanding of schooling and his clever approach to data analysis.

– **Bill Schmidt**

We also write this book to clarify the possibility for all children to experience both excellence and equality in their schooling. This includes those closest to us.

Keara	Ava
Frederick	Joella
Zoe	Norah
Shane	Grace
Arnold	Wesley
Carolyn	Audrey
Jaclynne	Ashley
Dylyn	Elayna
Reegyn	Jackson
Maegan	

Contents

List of Figures	*page* ix
List of Tables	xi
Series Editor's Foreword	xiii
Preface	xv
Acknowledgments	xvii
List of Abbreviations	xviii

PART I THE HISTORICAL DEVELOPMENT OF MODERN INTERNATIONAL COMPARATIVE ASSESSMENTS — 1

1. Beginning the Modern Investigation of the Role of Schooling across the Globe — 3
2. The Arrival of TIMSS and PISA — 20

PART II CONDUCTING INTERNATIONAL ASSESSMENTS IN MATHEMATICS AND SCIENCE — 43

3. Who Participates in International Assessments? — 45
4. What Students Know: From Items to Total Scaled Scores — 65
5. Relating Assessment to OTL: Domain-Sensitive Testing — 86
6. The Evolution of the Concept of Opportunity to Learn — 100
7. The 1995 TIMSS Curriculum Analysis and Beyond — 121
8. Characterizing Student Home and Family Background — 156

PART III THE LESSONS LEARNED FROM INTERNATIONAL ASSESSMENTS OF MATHEMATICS AND SCIENCE 181

9 Pitfalls and Challenges 183

10 What Has Been Learned about the Role of Schooling: The Interplay of SES, OTL, and Performance 210

11 Where Do We Go from Here? 239

Appendix A Third International Mathematics and Science Study 1995: Mathematics and Science Content Frameworks – Measuring Curricular Elements 255

Appendix B Third International Mathematics and Science Study 1995: Teacher (Implemented Curriculum) OTL Questionnaire for Grade 8 262

Appendix C Programme for International Student Assessment 2012: Student (Implemented Curriculum) OTL Questionnaire 278

Appendix D Trends in International Mathematics and Science Study 2015: Teacher (Implemented Curriculum) OTL Questionnaire for Grade 8 and Country Expert (Intended Curriculum) OTL Questionnaire 281

References 295

Index 310

Figures

1.1 The field of "comparative education": a brief history from approximately eighth century BC to 1969   *page* 5
1.2 Five historical phases of the IEA focusing on mathematics, revealing international assessment organizational shifts over time   18
2.1 IEA Tripartite Curriculum Model   25
2.2 Textbooks relation to curricular dimensions   26
2.3 Conceptual Model of Educational Opportunity   27
4.1 A Model of Mathematical Literacy in practice   68
5.1 Distribution of country ranks across mathematics subtopics   96
6.1 Carroll's Model of School Learning   107
6.2 The quadratic relationship of Applied Mathematics to performance illustrated for four countries   119
7.1 The three aspects and major categories of the mathematics and science frameworks   125
7.2 Number of mathematics topics intended   129
7.3 Mathematics topics intended at each grade by top-achieving countries   140
7.4 Science topics intended at each grade by a majority of top-achieving countries   141
7.5 A hypothesized structure for curriculum   144
7.6 Estimated structural model of curriculum and achievement for Japan, the United States, and Slovenia (eighth-grade mathematics)   147
8.1 Direct and indirect SES inequalities   178
9.1 Sources of variance for the eighth grade International Mathematics score   195
10.1 Variation in PISA-2012 OTL both across and within countries   215
10.2 Most of the variation in PISA-2012 OTL was within school   215

10.3	Model of SES inequality	221
10.4	Direct and indirect SES inequalities	224
10.5	Relationship across countries of size of OTL gap and performance gap at the between-school level	225
10.6	PISA-2012 predicted by TIMSS-2011 at the school level in Russia	231
10.7	Model for Russia relating OTL and eighth-grade performance to ninth-grade performance	234
11.1	Predicting PISA-2015 from TRENDS-2015	242
11.2	TIMSS and PISA schedule timeline for potential cohort longitudinal studies	245

Tables

3.1	Population definitions of IEA and OECD international mathematics and science studies	page 48
3.2	Percentage of sampled 15-year-old students at each grade level in the PISA-2012 study: OECD countries	52
3.3	Percentage of sampled 15-year-old students at each grade level in the PISA-2012 study: non-OECD participants	53
3.4	Participation in the PISA and IEA mathematics and science studies	58
4.1	Number of items addressing each mathematical content area for different student levels (populations) in FIMS	75
4.2	Subtest comparisons: SIMS population A (z-scores)	83
5.1	Average percent correct on country-defined sets of appropriate eighth-grade items	94
6.1	OTL measurement in IEA studies prior to the 1995 TIMSS	110
6.2	Average significant estimates and percentage of PISA countries with statistically significant relationships of OTL to PISA performance for each of the three measures of OTL	117
7.1	The share of textbook emphasis for each topic at grade 4	133
7.2	Percentage of mathematics textbooks that require a specific number of topics to cover 80 percent of the textbook	134
7.3	Percentage of science textbooks that require a specific number of topics to cover 80 percent of the textbook	135
7.4	Two representative topic trace maps for a selected set of countries	137
7.5	International Grade Placement (IGP) values for selected topics	138
7.6	Regression analyses relating coherence and focus to achievement	143

7.7	Examples of OTL measurement in IEA studies from 1999 TIMSS-R and TRENDS	149
8.1	TIMSS-95 student background items	170
8.2	SES related items from TIMSS-95/TRENDS and PISA student background questionnaires	171
8.3	Student wealth items included in the PISA-2012 SES index (ESCS)	173
8.4	Component loadings, loading ranks, and reliability for the ESCS composite from PISA-2012	174
9.1	Countries ranked in the top five in at least one international mathematics assessment	186
9.2	Countries ranked in the top five in at least one international science assessment	187
9.3	Means, standard deviations, and sample sizes for schools and classes by type of school and class track	200
9.4	Variation in OTL in non-tracked schools	201
9.5	Variation in OTL in tracked schools	202
9.6	Classroom means for four patterns of mathematics tracks, including their appropriate seventh-grade feeder class, and eighth-grade gain	203
9.7	Variation in mathematics performance in schools having no tracks	205
9.8	Variation in mathematics performance score in schools having tracks	205
9.9	The relationship of tracking and other student, classroom, and school variables to eighth-grade mathematics achievement in tracked schools	207
10.1	Sampling of journal articles using TIMSS/TRENDS and PISA data since 1995	212
10.2	Student-level correlations of composite SES measures with performance within each country for TIMSS-95, TRENDS-2011, and PISA-2012	222
10.3	Eleven countries with higher PISA mathematics scores and lower inequality	227
10.4	Estimated student achievement, PISA-2012, including TRENDS-2011 mathematics score	232

Series Editor's Foreword

In the last several decades, globalization has influenced the lives of all people. Business and education, as well as scientific disciplines, have all experienced the need to understand and work with people whose political, social, cultural, and linguistic origins are often very different. This has been true of psychology, education, and other social science disciplines. These developments also have important implications for the development and use of measures of human individual differences. Business and educational institutions using tests and institutions interested in certifying or accrediting test users have all experienced the challenges and opportunities generated by increased globalization.

Recognizing the need for the education of psychometricians and users of tests, Jean Cardinet spearheaded the formation of the International Test Commission (ITC) in the late 1960s and early 1970s. It was formally established in 1978. Current members include scholars and institutions from most of the European and North American countries as well as some countries in the Middle and Far East, Africa, and South America.

The major goals of the ITC are the exchange of information among members and furthering cooperation on problems related to the construction, distribution, and use of psychological measures and diagnostic tools. To accomplish these goals, the ITC has initiated a number of educational activities. The ITC has also developed and published guidelines on quality control in scoring; test analysis and reporting of test scores; adapting tests for use in various linguistic and cultural contexts; test use in general; and computer-based and internet-delivered testing; as well as a test taker's guide to technology-based testing. The ITC publishes a journal, *International Journal of Testing*. This peer-reviewed journal seeks to publish papers of interest to a cross-disciplinary international audience in the area of testing and measurement. In 2016, the ITC led the effort to produce the *International Handbook of Testing and Assessment*.

In 2013, the ITC proposed to Cambridge University Press a series of books on issues related to the development and use of tests. The goal of the series is to advance theory, research, and practice in the areas of international testing and assessment in psychology, education, counseling, organizational behavior, human resource management, and related disciplines. This series seeks to explore topics in more depth than was possible in the *Handbook* or in any single volume. The series will explore the national and cultural idiosyncrasies of test use and how they affect the psychometric quality of assessments and the decisions made on the basis of those measures. As such, we hope the series will contribute to the quality of measurement, but that it will also facilitate the work of professionals who must use practices or measures with which they may be unfamiliar or adapt familiar measures to a local context. We have asked both ITC members and other scholars familiar with a topic and who are also familiar with the global situation related to various topics to be the editors and contributors to individual volumes.

We are especially pleased to see this series develop and are confident that the books in this series will contribute to the effectiveness of testing and assessment throughout the world. Certainly, this volume on the international measurement of student achievement and the opportunity to learn contains a wealth of information on how to conduct educational assessments born of the authors' experience conducting seventeen international studies of mathematics and science achievement as well as student background and opportunity. The volume is a detailed examination of the role of schooling in these areas that spans twenty-five years. It includes chapters that address issues of student participation and how students' achievement, background, and learning opportunities have been measured in these studies. Seeing a compilation of this work by the outstanding scholars who have led these efforts in a single volume is truly exciting. We are hopeful that it will contribute to similar efforts in the future and that it will serve to enable educators to develop and use assessments to improve the education of their students.

We hope to publish a book at least biennially and encourage scholars who might be interested in developing a book proposal that addresses assessment in an international context to talk with the series editor, the ITC President, or other ITC leaders.

Neal Schmitt

Preface

Schooling matters. Our professional pursuits for more than twenty-five years have been focused on measuring one key aspect of schooling: the curriculum – what students are expected to study and what they spend their time studying. This we have done in an effort to document our conviction that schools and schooling play a vital and defining role in what students know and are able to do with respect to mathematics and science.

And yet, this assertion has been questioned, even today. Results from the extensive study of education conducted by Coleman and his colleagues in the early 1960s seemed to suggest that students' home and family background were far more important in determining what they knew as measured by academic assessments than their school experiences. If true, then policies aimed at ensuring equitable distribution of economic and other resources would be the important operative policy levers for improving overall student achievement.

However, another line of research, represented by the seventeen international studies of mathematics and science that are the focus of this book, provides a more nuanced perspective. While including multiple measures of student family and home background, these studies have also measured in multiple ways the substance of the learning opportunities students have had, and these have demonstrated a strong relationship with student performance. Furthermore, these studies have been able to document the interrelatedness of student background and learning opportunities. Student background matters to their learning opportunities and their achievement, yet their learning opportunities (schooling) matter even more.

This volume is both an extension and a continuation of our research examination of the role of schooling through international assessments of mathematics and science spanning the past twenty-five years. The chapters are organized in three major parts. Part I, "The Historical Development of Modern International Comparative Assessments," provides the historical, theoretical, and methodological context. Part II, "Conducting

International Assessments in Mathematics and Science," has chapters that delve into issues of participation and how students' achievement, background, and learning opportunities have been measured in these studies. Part III, "The Lessons Learned from International Assessments of Mathematics and Science," is our perspective on what has been learned from these studies together with a few concluding thoughts around the future for them.

Acknowledgments

We gratefully acknowledge the International Association for the Evaluation of Educational Achievement (IEA) and the Organisation for Economic Co-operation and Development (OECD) under whose auspices the seventeen studies that serve as the basis for this book were conducted. Their tireless attention to detail has provided the field with high quality data by which to examine the role of schooling worldwide.

It is also with deep appreciation that we acknowledge Richard Wolfe, who read the entire manuscript and provided a carefully thought-out set of comments. Jack Schwille was also kind enough to provide comments on the book.

Finally, it is with the utmost admiration and appreciation that we acknowledge our editor, Jennifer Cady. She worked tirelessly and meticulously over the last year to produce the drafts of the book. But, more than that, she reads with understanding, so she is able to catch not only errors of syntax and grammar but those of substance as well; for example, catching inconsistencies between tables and references in the text. Thank you, Jenn.

Abbreviations

ACER –	Australian Council for Educational Research
CASMIN –	Comparative Analysis of Social Mobility in Industrial Nations
CCSSM –	Common Core State Standards for Mathematics
CERI –	Center for Educational Research and Innovation
ESCS –	Economic, social, and cultural status
ETS –	Educational Testing Service
FIMS –	First International Mathematics Study
FISS –	First International Science Study
GDP –	Gross domestic product
IBE –	International Bureau of Education
ICCS –	International Civic and Citizenship Study
ICT –	Information and Communication Technology
IEA –	International Association for the Evaluation of Educational Achievement
IGP –	International Grade Placement
ILO –	International Labor Organization
INES –	Education Indicators Program
IRT –	Item Response Theory
ISCED –	International Standard Classification of Education
ISCO –	International Standard Classification of Occupations
ISEI –	International Socioeconomic Index
KMK –	Ständige Konferenz der Kultusminister der Länder
NAEP –	National Assessment of Educational Progress
NCES –	National Center for Educational Statistics
NCTM –	National Council of Teachers of Mathematics
OECD –	Organisation for Economic Co-operation and Development
OTL –	Opportunity to learn
PE –	performance expectations

List of Abbreviations

Pilot –	Pilot Twelve-Country Study
PIRLS –	Progress in International Reading Literacy Study
PISA –	Programme for International Student Assessment
SES –	Socioeconomic status
SIMS –	Second International Mathematics Study
SISS –	Second International Science Study
SMSO –	Survey of Mathematics and Science Opportunity
TALIS –	Teaching and Learning International Survey
TCMA –	Test Curriculum Match Analysis
TEDS-M –	Teacher Education and Development Study in Mathematics
TIMSS –	Third International Mathematics and Science Study
TIMSS-95 –	Third International Mathematics and Science Study
TIMSS-R –	Third International Mathematics and Science Study Repeat
TRENDS –	Trends in International Mathematics and Science Study
UIE –	UNESCO Institute of Education
UNESCO –	The United Nations Educational, Scientific and Cultural Organization
US –	United States
USOE –	US Office of Education

PART I

The Historical Development of Modern International Comparative Assessments

CHAPTER 1

Beginning the Modern Investigation of the Role of Schooling across the Globe

Schooling is ubiquitous across the globe. Virtually all of the almost 200 countries in the world have schooling as one of their central responsibilities. It is expected to prepare students to be responsible, informed citizens who will also contribute to the nation's economy. In that respect, schooling represents an investment in human resources that serves as a source of economic development as well as supporting the stability of society. Consequently, societies expect schooling to address both excellence and equality for their citizens. It is probably for this reason that around one-half of the world's countries have participated in one or more of the seventeen international assessments that have focused on mathematics, science, or both. Countries want some objective criteria on which to determine how well they have done on these two dimensions. This has led to the use of student achievement in these academic subjects, which have become ever more strongly related to productive and successful economies and that increasingly influence citizens' daily lives.

The metaphorical black box defining the key elements of schooling includes the student home and family background, which is brought to school; the content (including skills and reasoning) deemed by the society to be taught and learned; and the teacher who, with knowledge of the content and pedagogical skills, engages with the student over the content that is to be learned. To study schooling is to study these three key features.

A small group of university professors recognized that in order for comparisons to be made based on international assessment results, measures in addition to these assessments would be essential, since without them country comparisons would be meaningless. Using the black box metaphor, this led to the development and implementation of international measures of the home and family background of the student as well as measures that characterize what opportunities a student has had to learn appropriate academic content. These two factors – student socio-economic background (SES) and opportunities to learn (OTL) appropriate content – have been present in various instantiations in the international studies in a fairly consistent manner.

The issue of teacher quality, how well prepared a teacher is to understand the academic content and to prepare cogent and coherent lessons around this content as well as managing and maintaining a classroom environment conducive to student learning, is vitally important for student learning. Unfortunately, this is what has not been consistently measured in international studies and is often left out. International Association for the Evaluation of Education (IEA) studies have collected many different teacher background and instructional measures, yet perhaps due in part to the limits of the amount and kind of data that can be collected in cross-sectional studies, the types of measure included such as degrees earned, years of experience, and number of professional development activities attended are weak proxies for teacher quality.

As important as teachers are in schooling, few international assessment studies have developed meaningful measures of teacher quality that have demonstrated a relationship to student learning. One international study addressed this issue through an examination of teacher preparation that gathered an assessment of future teachers' knowledge of mathematics and related pedagogy at the end of their teacher preparation program. In this book we consider this groundbreaking international study of tertiary education as well as an additional study of the training, instructional practices, and beliefs of practicing teachers.

The focus, however, is on examining K–12 schooling across the globe by using seventeen assessment studies in mathematics and science that focus on OTL, SES, and student achievement as measured by curriculum-based tests and literacy tests.

This chapter provides the historical development of international assessments in mathematics and science in relation to the IEA from the formation of the organization and the first international education assessments through the Second International Mathematics Study (SIMS). This will allow the reader to position the IEA assessments in a historical context before moving to a greater understanding of the evolution of the international assessment stages that include the addition of the Organisation for Economic Co-operation and Development's (OECD) Programme for International Student Assessment (PISA).

Setting the Stage for the IEA

The origins of "comparative education" are mostly unknown, although the tradition can be traced as far back as the days of the Roman and Greek empires (Hans, 1949; Noah & Eckstein, 1969). Comparative education as a field has evolved since the Roman and Greek era, as have the theoretical and methodological underpinnings of the field. Arguably, the field has followed a similar developmental pattern expressed in the history of all comparative studies wherein "they all started by comparing the existing institutions. . . .. Gradually, however, these comparisons led the pioneers of these studies to look for common origins and the differentiation through historical development. It unavoidably resulted in an attempt to formulate some general principles underlying all variations" (Hans, 1949, p. 6). For the purposes of this chapter, we break down the development of the field of comparative education into five stages and provide a brief overview of what each stage entailed until reaching the point at which the IEA was formed. This will provide a clear lens with which to understand how assessment came to be viewed as an important tool in comparative education.

The five stages, adopted from Noah and Eckstein's book *Toward a Science of Comparative Education* (1969), as illustrated in Figure 1.1, move from general observations of foreign schools to borrowing aspects of schooling methods, finally developing into a field characterized by the

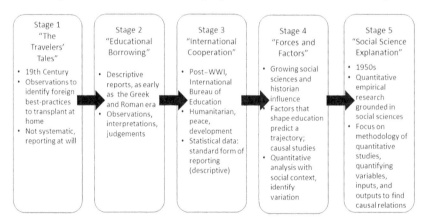

Figure 1.1 The field of "comparative education:" a brief history from approximately eighth century BC to 1969[1]

---

[1] The Stage titles in this model are directly quoted from Noah & Eckstein, 1969, pp. 4–7.

scientific method and its identification of variables and causal relationships between inputs (both within and outside the school system) and outputs of the school system. The first formal definition of the field has been credited to Marc-Antoine Jullien de Paris who, in 1817, saw comparative education as a way to analytically study education in other countries in order to modify and perfect national systems (Hans, 1949). Initially, the field (from a North American and European perspective) relied on descriptive comparisons that were used to help shape national education or to assess the values behind schooling while seeking best practices and eventually incorporating other educational philosophies (Cowen, 1996; Hans, 1949).

The first four stages increasingly incorporated the social sciences, building a foundation for the fifth stage of comparative education in the 1950s when the field sought deeper statistical understandings of various education systems. Following the mainstream argument that Sputnik and the Cold War altered the trajectory of the purpose of education, and therefore shifted the goal of comparative education, the move for comparative education to include more scientific comparative data through testing and statistics is not surprising (Cowen, 1996; Lundgren, 2011; Mitter, 1997). However, if we take a step back from the Sputnik theory and look at the history of the field we can see a gradual evolution into this stage, a trajectory that was already in place – Sputnik merely intensified the focus and movement in this direction.

Interest in the international comparison of education systems was already increasing around World War II when education began to be viewed as an investment in human resources and a source of economic development (Husén, 1967a). Political agendas increasingly focused on the need to identify scientific methods of comparison that served to highlight progress or areas of weakness as political competition between nations increased. The field of comparative education was a way to view the world guided by a lens of what is important in a specific context and at a specific time (Cowen, 1996). In 1949, eight years prior to Sputnik, comparative educationalists were calling for common statistics as a foundation for future comparison, including administration, organization, and tests of intelligence and achievement, at a time when "each country has its own terminology, based on national history, its own classification and its own method of collecting and compiling statistical tables" (Hans, 1949, p. 7). Such a situation made meaningful comparisons all but impossible.

While this is an incomplete picture of the state of the field at the time, it sets the stage for the 1958 research memorandum authored by Arthur Foshay of Teachers College, Columbia University, and sent to the United

Nations Educational, Scientific and Cultural Organization (UNESCO), which ultimately launched the initial phase of the development of the International Association for the Evaluation of Educational Achievement (IEA) (Husén, 1967a; Husén & Postlethwaite, 1996). The comparative education field is now replete with international assessments such as those found in the IEA studies, which adapt in purpose and structure to reflect a field that has become increasingly policy-oriented, increasingly competitive, and increasingly market-oriented (Broadfoot, 2010; Grek, 2009).

The Creation of the IEA

This fifth stage of "social science explanation" for comparative education sets the stage for the formation of the IEA and the launch of international comparative assessments in education that respond to the need for empirical evidence about student achievement and internationally comparative data. The consensus on which the IEA founding fathers based "the need to introduce into comparative educational studies established procedures of research and quantitative assessment" (Husén, 1967a, p. 13) was that previous research in comparative education provided only qualitative and descriptive information of education and culture, such as the descriptions from UNESCO, the International Bureau of Education (IBE), and the OECD, which were unable to provide insight into causal relationships among the educational inputs and outputs (Husén, 1967a). Before exploring the formation of the IEA, it would be pertinent to identify the founding fathers, together with their institutional affiliations, in order to better comprehend the context behind the conceptual framework of the IEA.

The IEA Founding Fathers

While there are many influential researchers who have been part of the IEA since the early days, there are a few who stand out as having taken an active and direct role in the actual formation of the organization. Often dubbed "the paternity," these founding fathers of the organization (listed here) are worthy of more space than this section allows:[2]

- C. Arnold Anderson, Professor, University of Chicago
- Benjamin Bloom, Professor, University of Chicago

---

[2] Founding fathers are listed with their academic position and affiliation held during the early days of the IEA and the First International Mathematics Study.

- Arthur W. Foshay, Professor, Teachers College, Columbia University
- Torsten Husén, Professor, University of Stockholm, Past Chairman and Technical Director of the IEA
- Douglas A. Pidgeon, National Foundation for Educational Research in England and Wales
- T. Neville Postlethwaite, Lecturer, St. Albany College of Further Education; Research Officer (Test Services), National Foundation for Educational Research in England and Wales, London
- Robert L. Thorndike, Professor, Teachers College, Columbia University
- W. D. Wall, National Foundation for Educational Research in England and Wales
- Richard Wolf, Graduate Student in Measurement, Evaluation and Statistical Analysis, University of Chicago – Studying under Professor Benjamin Bloom

The Formation of the IEA

To understand the full arc of the IEA and its impact on international assessments over time and in the present, it is important to know the goals and scope of the studies that began with the Pilot. Husén and Postlethwaite (1996) break the history of the IEA into five stages, whereas Gustafsson (2008) analyzes the organization in two stages. In this section, we will address the development of the first and early stages of the evolution and then in Chapter 2 we will provide clarity on the transition into the more mature stages and goals. As we will see in Chapter 2, the organization shifted into an administrative undertaking; however, in the early stages the IEA operated with a strong research orientation, where studies were built from research questions and hypotheses (Lundgren, 2011). The educational and social science researchers (the founding fathers) at this time in the late 1950s were interested in researching educational achievement and its determinants with the underlying assumption that factors influencing achievement are complex (Gustafsson & Rosen, 2014).

The unofficial start to the IEA was in 1958 when C. Arnold Anderson first addressed an interest in a comparative education research project that would help establish unified metrics for testing educational hypotheses. Following this came the memo by Foshay calling a variety of educational researchers to action. This action was a UNESCO Institute of Education (UIE) meeting in Hamburg followed by subsequent meetings in Eltham and other locales. The IEA was officially organized in 1959 with the aim "to look at achievement against a wide background of school, home, student and societal factors in order to use the world as an educational laboratory so as to instruct policy makers at all levels about alternatives in educational organization and practice" (Travers & Westbury, 1989, p. v).

Thus, as these founding fathers sought to fill a research gap with empirical evidence in a way that could easily be understood by specialists and nonspecialists alike, they focused on constructing instruments to evaluate problem areas related to school failure. The original purpose/mission of the IEA was to research education achievement and its determinants, to test hypotheses relating to educational outcomes based on social and cultural contexts, and to establish a "science of empirical comparative education" (Gustafsson & Rosen, 2014; Husén, 1967a, 1979, p. 371).

The initial IEA meetings were moved from UNESCO's UIE in Paris to their offices in Eltham and Hamburg as their Paris offices were too constraining for the organization. The IEA coordinating center moved to Stockholm in 1969 in order to accommodate a growing staff at the same time as the UIE's interest in the IEA waned. Furthermore, the Swedish University Chancellor was able to offer free computer time on the Ministry of Defense mainframe computer to the IEA while the data cleaning, weighting, and analyses were housed at the University of Chicago. Eventually, part of the data processing needed to move to the United States as part of a US Office of Education (USOE) grant requirement, at which time Thorndike established a data processing unit at Teacher's College, Columbia University. In this arrangement, data were cleaned and weighted with initial descriptive statistics at Teacher's College then sent to Stockholm for data analysis (Husén & Postlethwaite, 1996).

The Pilot Twelve-Country Study

As the IEA began to gain momentum following the initial meetings in Hamburg and Eltham, the first major (and logical) step was to test whether or not an international assessment of the type proposed would be feasible. This included identifying logistical issues that would need to be addressed prior to launching the first official assessment, and creating an assessment that is contextually appropriate and yields reliable data. The IEA's first international assessment was the Pilot Twelve-Country Study (Pilot). The stated goal of the study was to test various factors related to achievement, specifically "to be able to test the degree of universality of certain relationships which have been ascertained in one or two countries – for example, sex, home background, or urban-rural differences as related to achievement" (Husén, 1967a, p. 28). The group that had met in Hamburg to discuss the idea of the international assessment and form the IEA met three times between 1959 and 1961 in order to create the assessment, define the target student population, and plan the logistics of data collection and analysis (Foshay, Thorndike, Hotyat, Pidgeon, & Walker, 1962).

What sets the Pilot apart from the subsequent assessments is that it tested multiple areas (mathematics, reading comprehension, geography, science, and nonverbal ability), whereas the following assessments focused on single subjects until merging mathematics and science in 1995 (in the Third International Mathematics and Science Study).[3] The Pilot chose to focus on students aged thirteen years old, since that age group represented the final year that students would still be in school in all the participating countries, and ideally the samples were to be constructed based on students who were close to the national mean and standard deviation of achievement as understood in each country (Husén, 1967a).

The data were collected in 1960 and analyzed between 1961 and 1962, and from these analyses researchers presented the data in the form of national profiles. The main finding was that an international assessment of student achievement was feasible (Husén, 1967a). However, significant issues were recognized throughout the process, from the developmental stage to the analytical stages. The main issues related to data and expertise: Countries were not equally equipped with the skill set necessary to administer the test or to identify and conduct the appropriate sampling of schools and students, and the test items themselves had some translation issues. The key finding remains, however, that "what was most significant was that it proved that the project could be completed as planned" (Husén, 1967a, p. 29), so that the researchers at the IEA were encouraged to develop and launch the first formal international assessment of student achievement, the First International Mathematics Study.

First International Mathematics and Science Studies (FIMS and FISS)

As the Pilot demonstrated feasibility of such a large-scale international assessments of educational achievement, the IEA researchers were able to

---

[3] Even the Six Subject Survey, discussed on the following pages, broke the research into single-subject assessments rather than testing multiple subjects all during the same assessment time-frame.

turn their attention to a full scale, more complete study. The underlying sampling issues that were exposed during the Pilot were addressed as the IEA specifically contracted sampling experts in order to ensure the validity of their data. The First International Mathematics Study (FIMS) was developed in 1962 not to provide causal data related to academic achievement but instead to provide insight into how input factors (such as home environment, school procedures) relate to output measures (achievement); there was an underlying understanding among the IEA researchers that education was a social and political function, and FIMS sought to research how education responds to specific societal differences. In the development stages of FIMS, the researchers recognized and actively worked to address the limitations of the Pilot, which not only included the sampling errors and inconsistency with data collection but also included the need for a wider range of international actors to develop a mathematics assessment (Husén, 1967b, 1979; Schwille, 2011).

During this time, IEA remained a loose collaboration of researchers rather than a formal organization and it relied on the participating researchers and countries to raise the necessary funds for the studies, to identify hypotheses to study, to agree on and narrow these hypotheses into a manageable set of goals, and to collect and disseminate the data within their respective countries (Gustafsson & Rosen, 2014; Husén, 1979; Husén & Postlethwaite, 1996). In the era of FIMS, the purpose and studies of the IEA continued along a similar trajectory that revealed an emerging pattern of interest: While attempting to take account of how teaching and learning are influenced by developments in society, there was an interest in conducting longitudinal studies that were not feasible due to constraints of both time and resources.

The study was developed to be a scientific research project, rather than a simple statement of data; thus, even with limitations that prevented the IEA from developing a longitudinal study, the founding fathers improvised ways to maintain the integrity of the scientific study in order to produce data that would be an acceptable alternative. Thus, the researchers made the decision to test two different groups of students in four populations at a single point in time rather than developing a longitudinal study – an implicit cohort-longitudinal design.

The decision was made to focus on two terminal points in each educational system: the point at which nearly 100 percent of students of a particular age were still present in schools and the point immediately prior to university. The first terminal point group was associated with thirteen-year-olds, yet differences across the countries committed to participating in the study surfaced in identifying a single grade in which all of these students were enrolled. Consequently, the study defined two thirteen-year-old populations: population 1a was defined as all thirteen-year-olds in whatever grade they were enrolled, and population 1b was defined as the grade in which the majority of thirteen-year-olds were enrolled. The pre-university population consisted of two subpopulations: population 3a, defined as those students taking mathematics, and population 3b, those who were not.

The IEA researchers selected a single academic subject to test, mathematics, for three key reasons: First, while science and technology were identified as a pressing policy issue, mathematics was viewed as the foundation on which science and technology would be understood; second, there was a rise in recent efforts to reexamine mathematics curricula in schools in multiple countries, which suggested a need to examine the strengths and weaknesses of the curricula; and finally, mathematics was framed as a universal language that would minimize translation issues that were identified in the Pilot (Husén, 1967a, 1967b; Medrich & Griffith, 1992).

As a research organization, the IEA had multiple meetings during the test construction phase in order to identify hypotheses for independent variables worth testing as well as to define how to measure those variables. David Walker, Director of the Scottish Council for Research in Education, had proposed in the Pilot Study the "opportunity to learn" (OTL) measure as a method to understand the impact of teaching on learning in mathematics, based on the idea that what is taught by teachers is not always the same as what teachers claim to teach. The initial measure of the variable used in FIMS was crude, but it was the first time that this concept had been formally quantified and used as a variable to understand the relationship between inputs and outputs in educational achievement (Husén & Postlethwaite, 1996; McDonnell, 1995).

The assessment comprised multiple-choice items, a student opinion booklet, and background questionnaires given to students, teachers, and school experts in each country. The analysis conducted by the IEA produced a number of findings related to the impact of student attitude, gender, and socioeconomic status on achievement. The data were collected in 1964 and official reports published in 1967 (Husén, 1967a, 1967b; Postlethwaite, 1967). However, these official reports were followed by many case studies, technical reports, overviews, and other publications of national analyses of data that were published by the IEA researchers and other researchers in the participating countries (Husén, 1979; Husén & Postlethwaite, 1996; "IEA: Home," n.d.).

As with the Pilot and many other research projects, the completion of FIMS and the analysis of data produced not only results but also brought to light issues with the test and questions for future research. Translation, data collection, and the protracted timing of official reports were issues identified as problematic in the Pilot that persisted in FIMS, and some researchers questioned the generalizability of the study. Perhaps predictably, given the initial desire for longitudinal data, one of the critiques that arose after the publications were released was that the data only looked at one point in time and, therefore, was not able to provide insight into growth or changes within a system of education. However, similar to the response to the Pilot by the IEA, in which the researchers not only acknowledged the issues but worked to address them in the creation of the next study, these issues were not ignored or denied but rather brought to the forefront by those involved as topics to be addressed when future studies were considered.

In 1965, the researchers began to wonder whether or not the achievement predictors found in FIMS would be pertinent to other subject areas, which led to the Six Subject Survey that included the First International Science Study (FISS). Although working on six different assessments began to wear on the IEA and the international community at large, the Spencer Foundation Fellows were invited to Stockholm for secondary analyses of the FIMS and FISS data when the interest in international studies waned. This renewed interest encouraged the IEA to consider a second mathematics study (Gustafsson, 2008; Husén & Postlethwaite, 1996).

Second International Mathematics and Science Studies (SIMS and SISS)

Just as the IEA researchers did after the Pilot Twelve-Country Study, they recognized and made efforts to address the limitations of the previous study as they developed the next assessment. In particular, the IEA recognized the limitations of single-time studies and, short of transforming the studies into longitudinal studies, the IEA began to seek ways in which to address the issue of single-timepoint assessments. Following FIMS and the subsequent Six Subject Survey (which introduced FISS), the research climate began to shift slightly, which added another layer to the issues the IEA researchers needed to address.

The 1970s saw an increased interest in international large-scale assessments with a keen interest on quick results, while at the same time the "Cambridge Manifesto" (see Elliott & Kushner, 2007) highlighted the growing critiques of purely quantitative methods in educational research. The critique was that too little research was being directed to the actual teaching process, while too much attention was paid to student behavior due to the research climate focusing on precision in measurement (Lundgren, 2011).

Thus, mindful of the FIMS limitations and critiques of the field, the IEA developed the SIMS. The goals of SIMS were to contribute to a deeper understanding of education and the specific nature of teaching and learning. SIMS' intense focus on the context in which mathematics learning takes place was expressed through the emphasis on what happens inside the classroom, a new approach to studying teaching and learning, with a concern not only for what students learned but what the curriculum intended to teach and what was actually taught (Burstein, 1993; Husén & Postlethwaite, 1996; Travers & Westbury, 1989).

SIMS was specifically designed to include pieces of the FIMS assessment, creating the first possibility to compare/contrast with historical results among eleven repeat participating education systems. This allowed researchers to explore ways in which mathematics teaching and learning may have changed since FIMS (Travers & Westbury, 1989), and the researchers honed the Opportunity to Learn (OTL) measure that was piloted in FIMS. As SIMS had grown in scope as well as in the number of countries participating, the assessment was building legitimacy in its ability to contribute to a deeper understanding of education as well as to the overall nature of teaching and learning. Mathematics was again the chosen assessment subject, important due to the perception that mathematics was uniquely poised to broaden and hone intellectual capabilities, and would be used widely in life after school. As Travers and Westbury noted, mathematics "provides an exemplar of precise, abstract and elegant thought" (1989, p. 1).

SIMS introduced an optional longitudinal component to the assessment (in the lower secondary population), but IEA researchers also used items from FIMS in order to create some continuity and comparability across the two tests for those systems that chose not to participate in the longitudinal component. The IEA found itself in the midst of an interesting contradiction in education at this time, in the early 1980s. On the one hand, the late 1970s had shown a diminishing interest at the research level in large-scale international assessments, which was addressed by the Spencer Foundation's call for secondary analysis of the FIMS and the Six Subject Survey databases. On the other hand, national governments were increasingly interested in international comparisons and periodic studies with immediate results (Burstein, 1993; Husén & Postlethwaite, 1996).

Interest in periodic studies led to the Second International Science Study (SISS), yet the call for immediacy was not realized: The SIMS study was developed in 1976 and the test was administered in 1980–1981 but the three official publications were not released until 1987, 1989, and 1993. Similarly, the SISS study was administered in 1983–1984 but the official publications were released only in 1991–1992. At the same time, while SIMS and SISS took a while, the IEA researchers began to propose a multitude of other studies of interest both related and unrelated to the interest in periodic studies (Husén & Postlethwaite, 1996).

As the momentum of the IEA built, excitement was visible in the development of the SIMS assessment specifically related to the strong focus on teaching and learning. The OTL measure was refined, the optional longitudinal study was novel in international assessment, and the mathematics assessment broke down "mathematics" into smaller topic areas in order to understand achievement differences within mathematics at a deeper level. As the IEA looked beyond mathematics, the excitement was redirected to the larger concepts of teaching and learning:

> Yet [Jack] Schwille helped retain concern for the broader context that comparative and policy perspectives from IEA studies have so ably provided. It was always easy within SIMS to get caught up in the enthusiasm for the study of teaching and learning of mathematics *per se* and lose sight of the dual benefits of the worldwide "laboratory" of ideas and concerns available in an IEA conducted cross-national study (Burstein, 1993, p. xxviii).

Setting the Stage for the 1995 Third International Mathematics and Science Study (TIMSS-95)

The main issues that were brought to light through SIMS were twofold: the validity of comparisons and the rise of the "cognitive Olympics." Comparison was called into question in the absence of curriculum commonality among the participating countries, although the OTL measures sought to act as an adjustment that would enable achievement to be comparable in the absence of a common curriculum. While the OTL measure was significantly refined in order to serve that purpose, the concern is an important reminder that large-scale international assessments face a variety of hurdles related to test validity by the nature of the multitude of contexts in which the school systems are situated.

The question of validity and comparability, while addressed through the OTL in relation to curricular differences, continues to be an issue that researchers always need to address in such assessments (Bradburn, Haertel, Schwille, & Torney-Purta, 1991; Burstein, 1993; Husén & Postlethwaite, 1996).

The second issue was the rise of the cognitive Olympics, in which the data were being used to create rankings of the participating countries for political purposes (Burstein, 1993; Husén, 1979). This issue will not be as easily overcome as finding a method to control or adjust for unknown variables such as the OTL did for comparisons. Consequently, it will continue to be an issue pertinent to future assessments and should not go unnoticed. Despite the potential for the misuse of data, SIMS itself continued in the IEA tradition of building on past themes through identifying issues presented by FIMS and addressing them in the creation of SIMS. However, since the SIMS results and publications took such a long time, political shifts and international interests changed between reports and set the stage for the next phase in international comparative education: a shift away from research into evaluation that may be used in policy-relevant (or agenda-relevant) decisions (Burstein, 1993; Gustafsson, 2008; Husén & Postlethwaite, 1996). As the purpose in assessment shifted, the impact of SIMS remains strong:

> the early leaders were not so naïve as to think that wishing for equity made it so. Rather they were prescient enough to introduce what may be IEA's most powerful contribution of all to the literature on educational achievement surveys; namely, the measurement of opportunity to learn (OTL) (Burstein, 1993, p. xxxiii).

Paving the Way for the Next Phase

Research focusing on the historical development of international assessments of student achievement developed theories based on breaking the history of assessment into different stages; some research points to three macro-stages of international assessment, while other research suggests there are only two stages or as many as five stages. Of course, as time is continuous and international assessments continue to be conducted in the current era of international comparisons, it is to be expected that the number of stages in the evolution of international education assessment will increase.

This chapter and Chapter 2 have chosen to characterize the development of international assessment studies through five stages that merge the five stages proposed by Husén and Postlethwaite (1996) with the three stages of evolving assessment frameworks proposed by Gustafsson (2008) to create a new understanding of five phases/stages. The evolution of assessment frameworks and the tests themselves combine in a flow that is illustrated in Figure 1.2 as anchored by the mathematics studies (Gustafsson, 2008; Husén, 1979; Husén & Postlethwaite, 1996).

This chapter has sought to provide a historical overview and also to provide a context for the first stages in this model, which include the initiation of international achievement studies, the development of a framework for this type of research, and the work of the IEA researchers to learn from and build on past assessments rather than reinvent the wheel. Before moving into phase four, with the IEA introduction of the Third International Mathematics and Science Study (TIMSS-95) and finally into phase five, which brings another actor, the OECD, with another assessment, the Programme for International Student Assessment (PISA), it would behoove us to take a moment to answer the questions "what was learned by the start of the IEA?" and "under what context was the IEA pushed into TIMSS-95?"

The start of the IEA was instrumental in teaching educational researchers and social scientists that large-scale assessments of student achievement are feasible, albeit with many challenges that need to be addressed. As the IEA sought to learn from past assessments, not only in terms of the researchers' hypotheses but also in terms of technical and theoretical issues presented by the test itself, the IEA also serves as a lesson to researchers in general: research is not static; rather it is an action verb – something that is done, improved on, and changing. As SIMS showed, the trend to build on past experiences can provide researchers with unique and important measures such as OTL that can be refined in order to help address concerns about comparability.

At the same time, an ongoing struggle regarding the use of research and data is a reminder to researchers that even research projects that are planned, piloted, and reworked can be misused or misinterpreted. This is also an important lesson for policy makers, educators, and general consumers of research: to be critical of how data are presented and used. The political shifts and international interests briefly mentioned in the section about SIMS also provide a piece to this puzzle; phase four moves into the 1990s, an era of increased interest in knowledge societies, reemergence of human capital theories, and the end of the Cold War. This also marks the rise of the "Information Age," when data are expected instantly. As we will see, this will lead to increased pressure away from

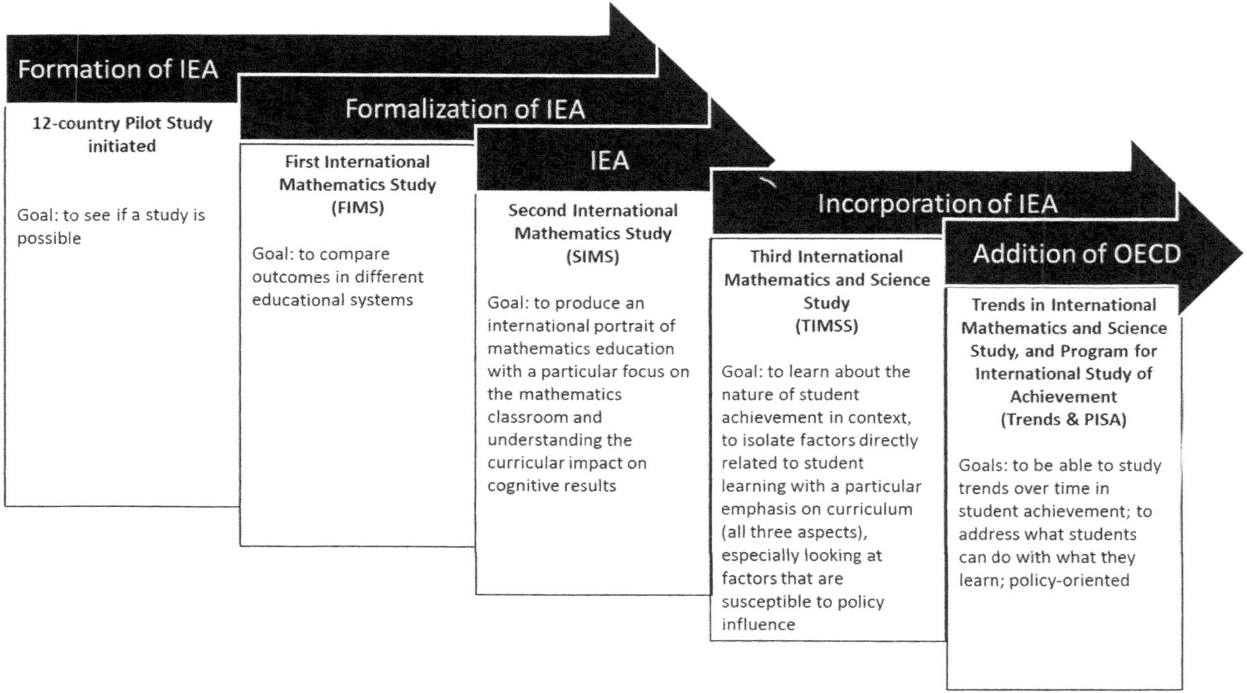

Formation of IEA

12-country Pilot Study initiated

Goal: to see if a study is possible

Formalization of IEA

First International Mathematics Study (FIMS)

Goal: to compare outcomes in different educational systems

IEA

Second International Mathematics Study (SIMS)

Goal: to produce an international portrait of mathematics education with a particular focus on the mathematics classroom and understanding the curricular impact on cognitive results

Incorporation of IEA

Third International Mathematics and Science Study (TIMSS)

Goal: to learn about the nature of student achievement in context, to isolate factors directly related to student learning with a particular emphasis on curriculum (all three aspects), especially looking at factors that are susceptible to policy influence

Addition of OECD

Trends in International Mathematics and Science Study, and Program for International Study of Achievement (Trends & PISA)

Goals: to be able to study trends over time in student achievement; to address what students can do with what they learn; policy-oriented

Figure 1.2 Five historical phases of the IEA focusing on mathematics, revealing international assessment organizational shifts over time

a research-oriented approach to such studies into a bureaucratic "indicator" process (Husén & Postlethwaite, 1996; Lundgren, 2011; Pereyra, Kotthoff, & Cowen, 2011).

As Torsten Husén and Neville Postlethwaite (1996) conclude in their historical look at the IEA, future studies must continue to answer relevant questions as expressed in the original mission of the IEA, and the future of the IEA (and, we would argue, international assessments more broadly) will depend on the quality of answers the assessments can provide. It is fitting to move into the next phase of international assessment with that observation firmly in mind.

CHAPTER 2

The Arrival of TIMSS and PISA

As noted in the previous chapter, the development of international assessments followed a progression from the IEA's Pilot Twelve-Country Study through the First International Mathematic Study (FIMS) and the Six Subject Survey, which included the First International Science Study (FISS) to the Second International Mathematics Study and the Second International Science Study (SIMS and SISS). The IEA founding fathers initiated these assessments wanting to know whether international comparative assessments were even possible. After successfully conducting the Pilot Twelve-Country Study, more expansive research projects were developed that sought to understand the factors related to educational outcomes and how those relationships varied over countries. Each project iteration followed a strong tradition of learning from past challenges and building upon the existing foundation of international assessment experience. The completion of SIMS and SISS clearly marked the beginning of a paradigmatic shift in the field of international assessments (Gustafsson, 2008; Husén & Postlethwaite, 1996). Nonetheless, true to the tradition of IEA studies, in some ways the Third International Mathematics and Science Study (TIMSS-95) was like SIMS in that it was influenced by previous IEA assessments, most particularly by the innovative nature of SIMS.

However, TIMSS-95 departed in *major* ways from those earlier IEA studies, ushering in a major paradigm shift in international studies. It continues to be unique both in terms of the nature of the research design and in the sheer size and complexity of the study, even among more recent IEA projects, such as those done every four years under the same name – TIMSS –where the "T" now stands for "Trends" rather than "Third" (to avoid confusion we have referred to the original TIMSS as TIMSS-95 and will continue to do so. The other subsequent IEA studies will be referred to as TRENDS). Affirming the IEA emphasis on curriculum, the TIMSS-95 *Curriculum Frameworks for Mathematics and Science* noted that "[o]ne of the hallmarks of IEA studies, and certainly of the more recent ones, has been the recognition given to the importance of curriculum as a variable in explaining differences among national systems and in accounting for differences among student outcomes" (Robitaille, Schmidt, Raizen, McKnight, Britton, & Nicol, 1993, p. 11).

As a result, TIMSS-95 had three major foci: "an in-depth analysis of mathematics and science curricula, an investigation of instructional practices based mainly on teacher self-reported data, and an assessment of students' mastery of the curriculum as well as their attitudes and opinions" (Robitaille et al., 1993, p. 12). Additionally, TIMSS-95, as it was described in the introduction to the first United States (US) national government report, was "the largest, most comprehensive, and most rigorous international study of schools and students ever" with "data on half a million students from 41 countries" (US Department of Education, 1996, p. 3). The unique status of TIMSS-95 does not rest solely on its breadth of students, teachers, and countries. TIMSS-95 is unique "because of innovations in its design ... [including] analyses of textbooks and curricula, video-tapes [of classroom teaching in three countries], and ethnographic case studies [in the video-taped classroom countries]" (US Department of Education, 1996, p. 3).

Informing and connecting all these various research components was an expansion of the IEA Tripartite Curriculum Model developed in SIMS (Travers & Westbury, 1989). This expanded conceptual framework along with the TIMSS-95 *Curriculum Frameworks for Mathematics and Science* (see Appendix A) served to link, both conceptually and methodologically, all aspects of the study as well as highlighting the notion that each education system is embedded in its own particular cultural and political context. Although some view TIMSS-95 as a shift in the IEA emphasis from *understanding* the factors related to country differences in performance to merely *describing* educational outcomes (e.g. Gustafsson, 2008), this characterization is entirely unwarranted and misleading, and is contradicted both by the above quotes found in the official US National Center for Education Statistics report and the nature of the resultant types of data and analyses that were produced in TIMSS-95 (Schmidt, McKnight, Houang, Wang, Wiley, Cogan, & Wolfe, 2001).

In this chapter, we will explore the paradigm shift in international comparative research embodied by the Third International Mathematics and Science Study (TIMSS-95). We conclude this chapter with an examination of the subsequent TRENDS studies and the rise of another international research organization, the OECD, and its respective international assessment, PISA.

From SIMS to TIMSS: Setting the Stage for the Next Phase of International Assessments

While it is tempting to start this chapter by simply looking at the Third International Mathematics and Science Study, we need to step back and look at how the atmosphere changed after the administration of SIMS in order to get a better understanding of how and why TIMSS-95 developed as it did. Recall that SIMS operated during a shift in international research agendas: after the 1970s, interest in large-scale education research and evaluation waned while at the same time governments became increasingly interested in international *comparisons* (Burstein, 1993). Furthermore, governments were interested in *immediate* results that they could use to compare their education systems internationally, thus highlighting the growing interest in a "cognitive Olympics" (Burstein, 1993).

The political situation surrounding SIMS and the future of the IEA was complicated by the challenge of elapsed time: SIMS took a very long time to plan, conduct, and particularly to produce reports. The SIMS research project was proposed in 1974, developed between 1974 and 1980, SIMS testing occurred between 1980 and 1982, yet initial analysis and reporting lasted from 1983 to 1987 (Travers & Westbury, 1989). The final official SIMS report was released in 1993 – about ten years after the data were collected and nearly twenty years after its conception (Burstein, 1993).

Between reports, there were political shifts and changing international views on education and the role of international research. Different political agendas proposed different research goals for the IEA ventures, ranging from the desire to learn from other education systems broadly to each system comparing only what is deemed important at the time (Burstein, 1993). Couple the timeline challenges of SIMS and the changing research agendas of participating countries with the fact that SIMS was only part of the larger work being done by the IEA at that time, including SISS, and recall that each participating country financed their participation in the project, and you begin to get a picture of how complicated the issues were for the IEA during the time of SIMS.

IEA researchers "were exhausted and frustrated" (Gustafsson, 2008, p. 2), under-funded, faced with design issues, felt pressured to speed up the research process, and were increasingly concerned that the studies were not adequately answering questions about the causes of educational achievement (Gustafsson, 2008; Husén & Postlethwaite, 1996; Schwille, 2011).

The challenges in international education research embodied by the SIMS' wrap-up also marked a shift in the IEA organization. In the late

1980s to early 1990s, "a new organization of IEA was set up, with a permanent secretariat in the Netherlands, and a data-processing centre in Hamburg" (Gustafsson, 2008, p. 2). The Six-Subject Survey had stretched the capacity of the IEA, and though the project was completed it was not repeated due to this issue; further, the IEA emerged from the study with a more formal organization that limited the ability of the IEA to continue to build capacity and shifted the organization away from its foundation as a community of scholars (Schwille, 2011).

However, under this new organization, which emerged toward the end of the SIMS/SISS-era, the IEA produced a template of quality standards and the goals were to begin to repeat studies that would allow for self-financing, capacity building, and would put their extensive institutional memory to use (Schwille, 2011). At the end of the 1980s, the IEA General Assembly proposed a third mathematics study, which subsequently was expanded to include science as the Third International Mathematics and Science Study with some forty participating systems of education (Husén & Postlethwaite, 1996). And thus, with the TIMSS-95, the era of TIMSS was launched.

The Road Leading up to the Third International Mathematics and Science Study

Not only was the Third International Mathematics and Science Study expanded from its initial conception as an eighth-grade study of mathematics to a mathematics *and* science study, under the influence of and the funding provided by the US government, it became a study of three populations: nine-year-olds, thirteen-year-olds, and students in their last year of secondary education. In essence it moved from a relatively small study of one content area in one grade to a quite complex design that furthermore included almost twice as many countries.

At the outset, the lead IEA researchers (including Leigh Burstein – University of California, Los Angeles, Senta Raizen – The National Center for Improving Science Education, David Robitaille – University of British Columbia, William Schmidt – Michigan State University, David Wiley – Northwestern University, and Richard Wolfe – University of Toronto) wanted to develop new measures of opportunity to learn (OTL) building on the work of SIMS. A grant from the National Science Foundation in the United States was awarded to Michigan State University in 1991 to develop the conceptual framework for the study and the new methodology for measuring OTL. The study was named the Survey of Mathematics and Science Opportunity (SMSO). The work of this project was a central part of the TIMSS-95 study and was integrated into the TIMSS-95 throughout its development.

More specifically, SMSO was to design the principal, school, and student questionnaires for TIMSS-95, including the methodology needed for the curriculum project in order to analyze the curriculum documents and textbooks of the participating countries, as well as to create the test blueprints based on early results from the curriculum analysis (Cogan & Schmidt, 2015; Schmidt, Jorde, et al., 1996; Schmidt & McKnight, 1995).

To accomplish these tasks, the SMSO included six countries (Spain, Japan, Switzerland, France, Norway, and the United States) as part of the research; with the first goal being the design and development of curricular content frameworks for both mathematics and science that would provide the linkage between the different TIMSS-95 components through a common language system (Garden & Orpwood, 1996; Schmidt & McKnight, 1995; Survey of Mathematics and Science Opportunities, 1993).

That work produced a common coding language with respect to mathematics and science content so that **all** TIMSS-95 components (curriculum document analyses, questionnaires, and the test blueprints) could be conceptually linked to produce the means by which each component of TIMSS-95 could be statistically related with *any other* component but especially to the assessment results – see Chapter 7 (Robitaille, Schmidt, et al., 1993; Schmidt, Jorde, et al., 1996; Schmidt, McKnight, Valverde, Houang, & Wiley, 1997; Schmidt, Raizen, Britton, Bianchi, & Wolfe, 1997). As part of the content frameworks the coding included different kinds and levels of expectations for student performance rather than relying on inferences as to the cognitive activity required to answer questions on the assessments (Schmidt & McKnight, 1995). The resulting framework (see Appendix A) became central to the TIMSS-95 curriculum analysis project, with the goal of developing multiple measures of opportunity to learn, each characterizing different aspects of the content coverage to which students were exposed.

The Third International Mathematics and Science Study Conceptual Model of Educational Opportunity

SMSO based the work of developing the TIMSS-95 conceptual framework on the notion that there are different aspects of the curriculum that impact opportunity to learn such as the intended and implemented curriculum, which were part of the model developed in SIMS (see Figure 2.1) (Burstein, 1993; Husén & Postlethwaite, 1996;

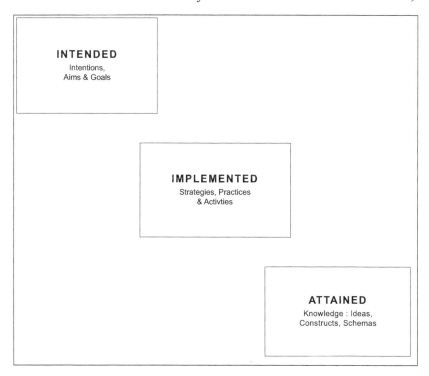

Figure 2.1 IEA Tripartite Curriculum Model
Note. Republished with permission of Springer Science and Business Media BV, from "Characterizing pedagogical flow: An investigation of mathematics and science teaching in six countries," by W. H. Schmidt, et al., 1996, p. 17; and from "According to the book: Using TIMSS to investigate the translation of policy into practice through the world of textbooks," by G. A. Valverde, L. J. Bianchi, R. G. Wolfe, W. H. Schmidt, & R. T. Houang, 2002, p. 5; permissions conveyed through Copyright Clearance Center, Inc.

Travers & Westbury, 1989; Valverde, Bianchi, Wolfe, Schmidt, & Houang, 2002). Further conceptual work extended the model to include textbooks as the potentially implemented curriculum – see Figure 2.2. SMSO then developed the TIMSS-95 Conceptual Model for Educational Opportunity (see Figure 2.3), a model of curricular intentionality and educational opportunity (Schmidt & Cogan, 1996; Schmidt & McKnight, 1995; Valverde, Bianchi, Wolfe, Schmidt, & Houang, 2002).

Figures 2.2 and 2.3 represent two interrelated models of curriculum related opportunities to learn used in the development of TIMSS-95. Curriculum was defined as an ordering and structuring of *student experiences* that are often specified by curriculum guides and textbooks.

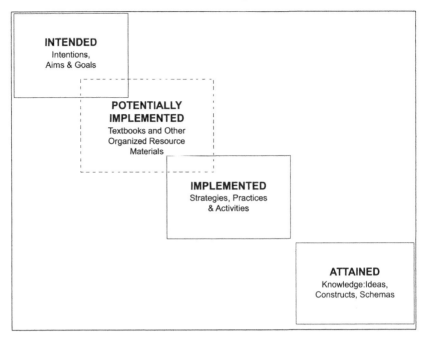

Figure 2.2 Textbooks relation to curricular dimensions

Note. Republished with permission of TIMSS Science, from *The Third International Mathematics and Science Study Technical Report: Volume 1: Design and Development*, by W. H. Schmidt & L. S. Cogan, 1996, pp. 5–8; and with permission of Springer Science and Bus Media BV, from "Characterizing pedagogical flow: An investigation of mathematics and science teaching in six countries," by W. H. Schmidt, et al., 1996, p. 30; and from *Many Visions, Many Aims: A Cross-National Investigation of curricular Intentions in School Mathematics*, by W. H. Schmidt, Curtis C. McKnight, Gilbert A. Valverde, R. T. Houang, & David E. Wiley, 1997, p. 178; permissions conveyed through TIMSS Science and Copyright Clearance Center, Inc.

This produces a guided distribution of opportunity, but teachers are the ones who further shape and define such opportunities in greater detail so that then they become the *implemented* curriculum. What is specified in curriculum guides defines the *intended* curriculum, in other words, what the student is expected to learn as detailed by the country, region, or school. The opportunities provided in textbooks or other instructional materials are only *potential* as the teachers who shape opportunity in detail may choose to what extent, if any, to follow the textbook.

Finally, there is whatever the teacher – using guides, textbooks, and other materials – decides to teach, in what order, and for how long. It, in

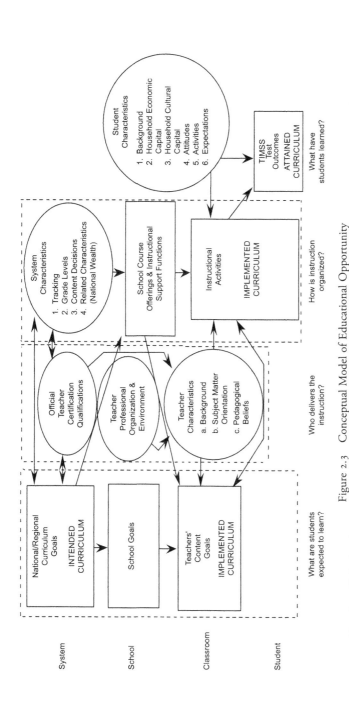

Figure 2.3 Conceptual Model of Educational Opportunity

Note. Republished with permission of Springer Science and Business Media BV, from "Characterizing pedagogical flow: An investigation of mathematics and science teaching in six countries," by W. H. Schmidt, et al., 1996, p. 19; and from "According to the book: Using TIMSS to investigate the translation of policy into practice through the world of textbooks," by G. A. Valverde, L. J. Bianchi, R. G. Wolfe, W. H. Schmidt, & R. T. Houang, 2002, p. 7; permissions conveyed through Copyright Clearance Center, Inc.

the end, defines what opportunities students actually experience as noted in Figure 2.3. The *attained* curriculum is what opportunities actually turn into learning, typically as measured by an assessment.

The *intended* and *potentially* implemented curriculum components were the key elements to be analyzed in the TIMSS-95 curriculum analysis. This produced both qualitative and quantitative measures of OTL, which, when combined with measures of the *implemented* curriculum from teacher questionnaires, provided a rich and deep characterization of students' opportunities to learn thus producing the extensive background the founding fathers saw as necessary to make sense of cross-country performance differences.

The Conceptual Model of Educational Opportunities (see Figure 2.3) represents the SMSO characterization of how curricular intentions influence classroom activities and addresses the complex set of factors that influence how curriculum interacts with student attainment given student characteristics, school characteristics, teacher characteristics, and learning expectations (Schmidt & McKnight, 1995; Schmidt, Jorde et al., 1996; Schmidt, McKnight, Valverde, Houang, & Wiley, 1997; Schmidt, Raizen, Britton, Bianchi, & Wolfe, 1997).

This particular model became the conceptual model, not only of SMSO but of TIMSS-95 as well (Schmidt & Cogan, 1996). It captures a variety of aspects related to the shaping of educational opportunities, even factors beyond the intended, potentially implemented, and implemented curriculum, in order to address the complexity of the four-part model of the curriculum (Schmidt, McKnight, Valverde, et al., 1997). In other words, this model adds detail to each of the curricular components in Figure 2.3, thus addressing differences in the attained curriculum within a country/system/school/classroom. The model of Figure 2.3 guided the development of the research design as well as the development of the measurement procedures and instruments used throughout the study.

The Design, Execution, and Reporting of the Third International Mathematics and Science Study

Based on the conceptual model relating OTL and other factors to achievement and the attendant data collection methodologies, the TIMSS-95 launched a new era of IEA research. Recognizing the desire of the IEA to study the determinants of academic learning and how SIMS addressed this with a longitudinal sub-study, TIMSS-95 introduced the idea of using adjacent grades in the study to measure achievement growth between grades (Gustafsson & Rosen, 2014; Husén & Postlethwaite,

1996; Wiley & Wolfe, 1992). It was also the first international study to take extensive advantage of new methodologies in assessment such as Item Response Theory (IRT). Other methodologies such as multiple assessment forms, partial replicate designs, matrix-sampling, and more rigorous standards guiding sampling, data collection, and timelines were also included.

Furthermore TIMSS-95 was based on a conceptual model in which the measurement of opportunity to learn and its relationship to performance was the central defining element. The development of detailed content frameworks provided the metric by which this could be achieved. Content covered in textbooks and national curriculum standards, teacher indicators of what topics in mathematics and science were implemented in their classrooms, together with the item-by-item content being tested on the assessment were all linkable through the common framework and its associated coding. The driving research questions were about understanding the relationship of OTL to achievement and learning and how it varied across countries as well as how such opportunity was distributed both across and within countries (Schmidt et al., 2001). These questions, the models in Figures 2.1–2.3, and the instrumentation described in Chapter 7 were all developed around this key issue and are reflective of the IEA's tradition and its very origins of looking at how opportunity to learn (i.e. the curriculum) is related to student performance.

The TIMSS-95 questionnaires, content frameworks, curriculum study methodologies, and test blueprints were developed by researchers working on the SMSO project in conjunction with TIMSS-95 staff from the international center. The development of the items and the assessment forms, as well as the actual field execution of the study, the IRT scaling of the assessments, and quality control, were carried out by the international center at Boston College in conjunction with the Australian Council for Educational Research (ACER).

Subsequent to the collection of the data and the scaling of the assessments, separate international reports were produced for both mathematics and science at each of the three populations (fourth grade, eighth grade, and end of secondary for most countries) by the international center at Boston College. All of these reports can be found online on the IEA website (https://timssandpirls.bc.edu/timss1995i/TIMSSPublications.html). Individual countries had the option of producing country-specific reports, which many countries did, including the United States, Germany, and Japan. Four books that focused on the cross-national comparisons of curriculum were also produced by the lead researchers who were also part of the SMSO team that designed many aspects of the study (see reference section for additional information):

- *Many Visions Many Aims: A Cross-National Investigation of Curricular Intentions in School Mathematics,*
- *Many Visions Many Aims: A Cross-National Investigation of Curricular Intentions in School Science,*
- *Facing the Consequences: Using TIMSS for a Closer Look at U.S. Mathematics and Science Education,*
- *According to the Book: Using TIMSS to Investigate the Translation of Policy into Practice through the World of Textbooks.*

A book analyzing the structure of curriculum (OTL) across thirty-two countries and its relationship both to achievement and to gains in performance – *Why Schools Matter: A Cross-National Comparison of Curriculum and Learning* – was also produced.

Some have suggested that this movement toward more sophisticated methodologies with their more demanding requirements together with greater accountability to government interests has lessened the research aspect of IEA studies but this was not the case, at least with respect to the original Third International Mathematics and Science Study (Gustafsson & Rosen, 2014). Given the research orientation of the books cited in the previous paragraph, this observation is clearly not characteristic of the TIMSS-95 but the characterization may be more relevant to the subsequent TRENDS studies that followed the original TIMSS-95 study.

TIMSS-95 was also the product of a changing IEA. Organized under the auspices of the IEA, which was headquartered in The Hague, in the Netherlands, the TIMSS International Study Center, under the leadership of Albert Beaton, was located in the Lynch School of Education at Boston College (now the TIMSS and PIRLS International Study Center). The IEA had established a Data Processing Center in Hamburg, Germany, to handle data analysis and cleaning. Additional expertise and advice around technical issues such as sampling and the psychometric scaling of the tests was provided by people from other institutions, e.g. Pierre Foy, Statistics Canada; Ray Adams, Australian Council for Educational Research (ACER), and Andreas Schleicher, Hamburg University. Item development was provided in part by the Educational Testing Service in the United States and ACER (Howie & Plomp, 2006). What this is to say is that the IEA, which began and functioned for thirty years as a loose confederation of university researchers, had become a more formal and structured entity that worked with countries according to a more precisely articulated agenda and timeline for conducting high quality studies and producing timely reports.

As mentioned multiple times over the course of this chapter, TIMSS-95 represented a major paradigm shift in the IEA's technical approach and its increased focus on rigor and timelines in international assessments, but it also represented a broadening in terms of research approaches, scope, and methods. TIMSS-95 used a variety of assessment types, including short answer and extended responses on the main assessments as well as a separate performance assessment that entailed the use of a complex scoring rubric. Both mathematics and science were assessed with some test questions geared toward the integration of the two areas. Students were sampled and assessed at three different age/grade stages of their schooling: nine-year-olds, thirteen-year-olds, and those in their final year of secondary. To support the cohort-longitudinal (synthetic) analyses for the nine- and thirteen-year-old population students in the grade immediately prior to the focal grades were also sampled and assessed. Finally, the end-of-secondary population had three subpopulations: mathematics specialists, science specialists, and generalists, which was defined to include all students at the end of schooling including the two specialist subpopulations.

Video Study

The TIMSS-95 project included lead-up studies such as the SMSO project and concurrent studies such as the videotape study (Stigler, Gonzales, Kawanaka, Knoll, & Serrano, 1999). The purpose of the videotape study was to collect information on eighth-grade instructional practices by going directly to the source (specifically by videotaping classroom instruction). The underlying theory in this study was that instructional processes are related to student learning (Hiebert & Stigler, 2000; Stigler & Hiebert, 1997). The videotape study focused on Germany Japan, and the United States and on eighth-grade mathematics only. While there were challenges, notably due to the labor-intensive approach, methodological challenges such as sampling and coding, the visuals this study produced added to the descriptive understanding of teaching as a cultural practice that could be characterized by quantifiable indicators to limit visual or cultural bias in the findings (Stigler & Hiebert, 1997).

What is significant about this study is that it was conducted by taking a random subsample of 231 eighth-grade classrooms from the TIMSS-95 sampled classrooms in the three countries – 100 in Germany, 81 in the United States, and 50 in Japan, thus highlighting TIMSS-95's strong research focus on instructional experiences as part of the curriculum model of Figure 2.3 (Stigler, & Hiebert, 1997; Stigler, Gonzales, Kawanaka, Knoll, & Serrano, 1999). The sampling design for the video study permitted analyses relating the variables created in the coding of the videos to OTL, SES, performance on the assessment, and other TIMSS-95 variables in a way that is generalizable to the eighth-grade population in each of the three countries.

Secondary School (Generalists) Study

Two other aspects of TIMSS-95 contributed even further to the extensiveness of the study and its enormous complexity. The first was the study of achievement for students in their last year of secondary school. This aspect of the study, however, never received the same amount of visibility that the other two grade levels did. There were two main reasons for this; unlike the primary and lower-secondary part of the TIMSS-95 study, the definition of secondary school was more complicated within many countries and across countries as well. Many countries themselves had different types of secondary schools, e.g. academic, vocational, technical, while others had a comprehensive system where all types of programs were found in the same school. In some countries this was further complicated by the fact that secondary schooling for some students included apprenticeship programs where students spent most if not all of their time in the work place rather than in a school setting. These differences made defining the population and the appropriate sample much more difficult for individual countries as well as interpreting the results across countries.

The second difficulty was comparability, as the number of years of schooling that accrued for end of secondary-school students varied across countries. The grade level ranged from grades 10 to 14, but for most countries end of secondary was twelfth grade. Even within countries, this could vary. For example, in Iceland the end of secondary schooling included grades 12, 13, and 14 depending on the school (US Department of Education, 1996).

Using country-specific definitions, students in their last year were sampled and all were given a mathematics and science literacy assessment similar in content level to that given at eighth grade. This was done because not all end-of-secondary-school students would be currently taking mathematics and/or science and what such students would have taken during their secondary schooling would have varied due to the common practice of tracking at the secondary school level in most countries.

Secondary School (Specialists) Study

Two subpopulations were identified: those who had taken advanced courses in mathematics and separately those who had taken advanced courses in science. In the United States the advanced mathematics students were defined as those taking pre-calculus or calculus. Given the organization of the mathematics curriculum in most other countries, such a single topic could not be used in determining the population but rather the designation of students taking advanced mathematics was used with the specifics left to each country. For science the same approach was followed, with the US course being college-preparatory physics and other countries again used the designation of advanced science courses taken.

For the two advanced populations, students took their relevant test focused on advanced content areas in mathematics or science. The mathematics test focused on calculus and the science test focused on physics. Twenty-two countries participated in the general literacy part of the study, while seventeen and eighteen countries took the advanced mathematics test and the physics test, respectively. No OTL data were collected relative to these students, and neither were there teacher data. This component of TIMSS-95 was viewed more as an indicator study rather than a research study as was the case for the other grades. An international report was prepared.

Performance Assessment Study

The other, and probably least well-known, component of TIMSS-95 was the study of performance assessment. This assessment was composed of thirteen performance tasks (twelve at each grade level): five in mathematics, six focusing on science, and two including tasks integrating mathematics and science content. Of the thirteen tasks, eleven were similar for both the fourth- and eighth-grade assessments with one unique task for each of the two grade levels. The tasks were chosen to represent the performance expectation aspect of the content framework (see Chapter 7 and Appendix A). An example of a performance task in science for eighth grade entailed work with batteries:

> In the Batteries task, students were provided with four unmarked batteries and a flashlight. To begin, they were asked to find out which of the batteries were good and which were worn out. The task was intended to measure students' ability to develop and implement problem-solving strategies and use experimental evidence to support a conclusion, but it also sampled

specific knowledge about electricity to solve a routine problem and to develop a concept-based explanation for the solution. Item 1 required students to identify the good batteries, which could be achieved by a systematic process of trial and error. Item 2 called for a description of the strategy used to identify the good batteries. Item 3 in this task required selection of the correct arrangement of batteries in a flashlight. Item 4 asked students to explain why their solution was correct, which requires knowledge of the concept of a complete circuit and an understanding of the direction of flow of electrical current (Harmon et al., 1997).

In mathematics an example at grade eight involved packaging:

> The Packaging task involved problem solving in three-dimensional space. Students were supplied with four small plastic balls packed into a square box, some sheets of light cardboard, and an explanation and illustration of a net for the box. With these, and a supply of materials such as a compass, ruler, scissors, adhesive tape, and paper clips, students were to find three other boxes in which the balls could be tightly packed, sketch the boxes, draw a net for each one, and then draw one of the nets to the actual size needed to hold the four balls. The task is intended to measure the students' sense of spatial relations as evident in their ability to visualize different arrangements of objects in boxes, to translate the three-dimensional models first into a two-dimensional sketch, then into the corresponding net, and finally to scale the net to actual size, working from concrete materials rather than by applying a formula to measurements (Harmon et al., 1997).

The performance assessment component was not required for participation in TIMSS-95, and as a consequence twenty-one countries participated – twenty-one countries at grade 8; ten of them also participated at the fourth grade. The samples for the countries were drawn as subsamples of the fourth- and eighth-grade samples that had participated in the main TIMSS-95 assessment. Similar to the original IEA Pilot Study's exploration of the possibility of a comparative assessment conducted in 1959–1962, the performance assessment component of TIMSS-95 was viewed as a pilot to see if performance assessments could be part of an international study.

The selection of the tasks to be included in the assessment proved to be difficult. With countries from different parts of the world living under different environment conditions, what might be of scientific interest to students in Norway could be very different from that found in Greece. The second major challenge was the development of the scoring rubric for each performance task.

A formal final report was issued by the International Center at Boston College (Harmon et al., 1997) and showed differences in performance across the countries for each of the individual performance tasks separately. No other special concurrent data were collected and no relational analyses were performed other than illustrating gender differences in performance. The report also focused on indicators of performance compared across the participating countries by providing average scores across the tasks to represent overall performance.

Case Studies Component

To put the various results of the TIMSS-95 study in a cultural context, case studies of the educational systems in each of the three countries that participated in the video study – Germany, Japan, and the United States – were created. The use of mixed methods was particularly appropriate in this case, as extensive statistical data were available from teachers, students, and principals. Additionally, classroom instruction was videotaped. The case studies were designed to bring all of these data and visual images into the educational and cultural context for each of these countries (see Stevenson, 1998; Stevenson & Nerison-Low, 1997).

TIMSS-Repeat and Beyond: Trends in Mathematics and Science Studies

The first IEA study to follow up on the Third International Mathematics and Science Study was called the Third International Mathematics and Science Study Repeat (TIMSS-R). The idea behind the study was to see what changes in performance had occurred over the four-year period since 1995, so as to provide trends in eighth-grade performance at the country level. Furthermore, the Population I sample in TIMSS-95 was now in eighth grade (or seventh grade for some countries), which became the sole focus of TIMSS-R. There was great interest among some countries but especially the United States to follow up on the US fourth graders who had performed relatively better than the eighth graders. Policy analysts wanted to see if that would lead to a better (in a relative sense) outcome at eighth grade for the United States four years later.

Thirty-eight countries participated in TIMSS-R, which was greatly reduced in its design complexity from the original TIMSS-95, for example eliminating the use of adjacent grades to provide a cohort-longitudinal study and focusing solely on eighth grade or its equivalent. It was again led by the International Center at Boston College. Unlike TIMSS-95, the United States lead on the study participation was directed by the National

Center for Education Statistics rather than any of the researchers who designed the original TIMSS-95 study. There was no curriculum study included but there was a new video study in seven countries (Gonzales, Calsyn, et al., 2000). Both mathematics and science eighth-grade classrooms (100 per country) were videotaped. Schools were the first stage sampling unit but the sampled classrooms for the video study were typically not the same as those sampled for the assessment.

Following TIMSS-95, the IEA studies became characterized by more country-level involvement and less university professor involvement. Previously, IEA assessments, up to and including TIMSS-95, were designed based on questions of interest to the educational researchers and the entire endeavor was viewed as a research study. Now the studies focused primarily on the routine administration of the assessment – in short, the studies became more bureaucratic and oriented around providing indicators for comparisons to previous assessments and for cross-country comparisons. Databases were developed and disseminated in order to allow researchers access to the data to conduct secondary analyses, which is another example of how TRENDS greatly regularized the work of the IEA.

TIMSS-R and subsequent TRENDS assessments view TIMSS-95 as the starting point to *periodic* assessment on a regular basis. From this point forward, TRENDS was administered at regular four-year intervals, enabling the monitoring of change at the same time as enabling countries to compare mathematics and science achievement to the students of other countries. The interest in *regular assessments* in order to monitor trends in performance marked a new goal: to accurately inform policy (Martin & Mullis, 2006). Regularized assessments also allowed countries to focus on their own education systems and changes in achievement alongside peer countries using opportunity models as a framework in order to limit the feeling of "horse races" (Schmidt & McKnight, 1995). The populations from which TRENDS samples were drawn remain grade-focused since they are assessing what students know *at a given point in schooling* (Cogan & Schmidt, 2015).

OECD and PISA

Around the same time that the IEA expanded and developed a periodic assessment, another recurring assessment of international education entered the scene: the Programme for International Student Assessment, or PISA. PISA is administered through the Organisation for Economic Co-operation and Development (OECD). PISA's stated purpose is to "evaluate education systems worldwide by testing the skills and knowledge of 15-year-old students" (OECD, n.d.), though more specifically PISA is aimed at measuring how well students can *apply* what they learn in school for use outside of school (OECD, 2000) and measures the flow of human capital from the schooling system (Sellar & Lingard, 2014). The OECD considers human capital to be composed of the knowledge, skills, competencies, and attributes of individuals that allow for the creation of well-being personally, socially, and economically (Sellar & Lingard, 2014).

OECD Background

The OECD began in 1961 as a reorganization of the former Organisation for European Economic Cooperation, which considered a wide variety of data collection and analysis as part of the core of its work, a demand that grew thanks to globalization (Sellar & Lingard, 2014). The OECD structure and expertise position is as a transgovernmental network for policy experts to meet and to devise responses to common challenges (Morgan & Shahjahan, 2014; Mundy & Ghali, 2009).

The OECD was not immune from larger political and economic changes around the world; countries were moving toward more decentralization, governance was more fragmented, curricula were diverse, and the OECD needed to govern by goals rather than methods. The fall of the Berlin Wall and the end of the Cold War changed international goals, and between-country competition shifted to economic power and growth rather than territory (Lundgren, 2011). In the post-Cold War era, the OECD underwent another rebranding and positioned itself as a center of "policy expertise and comparative international data" by creating programs of measurement, comparison, and analysis (Sellar & Lingard, 2014, p. 920).

History of the OECD Study of Education

PISA was not the OECD's first education-based project. In 1968, the OECD established a Center for Educational Research and Innovation (CERI), which was a policy institute (Lundgren, 2011). As competitive economic advantage was increasingly linked to education (Sellar & Lingard, 2014), the OECD considered education as a key part of their economic development mission. Therefore, education shifted from an informal to a formal component of OECD programs in 1975 when the Directorate for Social Affairs, Manpower and Education was created.

In 1991 the directorate was renamed the Directorate of Education, Employment, Labor and Social Affairs. In the 1980s, the OECD created the Education Indicators Program (INES) with the goal of building a system for education statistics that would enable comparisons across OECD countries (Lundgren, 2011; Morgan & Shahjahan, 2014). The INES program was tasked with regularly publishing key education indicators from OECD countries in the *Education at a Glance* publications. The 1990s drove the importance of education to the forefront of the OECD agenda with the advent of the "knowledge society," where human capital was based more on knowledge capabilities than traditional forms of labor, and thus the OECD released a policy position on lifelong learning and knowledge-based economies (Lundgren, 2011; Mundy & Ghali, 2009; Sellar & Lingard, 2014).

The 1990s also saw the United States advocate for OECD statistics and push for more streamlined and cohesive educational databases in order to develop measures of the potential global competitiveness of national economies, to promote policies encouraging growth of member economies, and to develop trade and the world economy (Lundgren, 2011; Mundy & Ghali, 2009; Sellar & Lingard, 2014). During this time, the quality of data and publications produced by the INES increased and the OECD was able to influence policies via annually published statistics with authority that was developed over time (Lundgren, 2011; Morgan & Shahjahan, 2014). The OECD General Assembly met every other year to develop the direction of the education program and sought clear support and mandates from member countries, which ultimately led to the introduction of PISA (Lundgren, 2011).

PISA History

> The rise of the OECD's education work is linked to the 'economization' of education policy and what we might see as the simultaneous 'educationizing' of economic policy, linked to the growing significance of skills and human capital agendas across the Organization (Sellar & Lingard, 2014, p. 921).

Since the 1970s, there has been a call for new research methods and models in educational research alongside international demand at the country level for more regular comparative data on the performance of education systems, particularly coming from the United States after the release of the 1983 *A Nation at Risk* report (Lundgren, 2011; Sellar & Lingard, 2014). The international pressure in conjunction with the re-emergence of human capital theories and the advent of knowledge societies in the knowledge economy not only set the stage for an increase in the OECD's interest in education but for the creation of PISA in particular.

The OECD set their sights on answering the international demand and planned to focus on outcomes and accountability to make education transparent, yet faced the immediate challenge of *how* to report learning outcomes (Lundgren, 2011). Apart from the INES annual indicators, the only comprehensive international educational data came from the IEA (Lundgren, 2011). The OECD attempted to incorporate the IEA data in the INES *Education at a Glance* publication but the IEA process was so lengthy that released publications were no longer relevant (Lundgren, 2011).

Further, according to OECD conception, the IEA did not match the need governments had as it did not provide comparable evidence of educational outcomes on a regular basis in such a way that it would be able to inform policy (Harlen, 2001). Not only did IEA studies not meet the need identified by OECD, the IEA had data issues since the tests and country participation changed over time, thus limiting the potential for international education comparisons (Lundgren, 2011). Although IEA results discounted the perceived relationship between education and the economy since the top performers were not the top economies, the OECD continued to focus on the role of education in economic development and sought to initiate its own international assessment (Harlen, 2001; Lundgren, 2011).

Thus came the idea of an international education study by the INES and CERI. Negotiations with the IEA were initiated but ultimately not pursued, PISA was developed in 1997, and the first PISA assessment was launched in 2000 (Harlen, 2001; Lundgren, 2011; Sellar & Lingard, 2014). The OECD started PISA to provide policy-oriented international comparable indicators of achievement regularly and over time, and to provide evaluation of educational equality explicitly for policy purposes (Gustafsson, 2008; Harlen, 2001). When taken in conjunction with IEA assessments and supported by policy communities, PISA provides a global infrastructure for international education comparisons that are particularly salient in an environment where policy focuses on numbers (Mundy & Ghali, 2009; Sellar & Lingard, 2014).

Creating PISA Goals and Framework

PISA takes a different approach to educational assessment than the IEA. Given the drastic and continuous changes in education policy discussions during the development of PISA, it focused on competencies rather than specific academic content and was informed by other assessments and the

Educational Testing Service (ETS) shifts in definition of "literacy" to include comprehension (Lundgren, 2011).

> Competencies in reading and in mathematics have to be continuously practiced. This means that the environment must offer possibilities to read and to calculate. The outcomes of PISA we hoped could stimulate a debate on learning outcomes not only from an educational perspective but also a broad cultural and social perspective (Lundgren, 2011, p. 27).

The specific domains included in PISA were identified based on what will be needed in adult life (Harlen, 2001). The "test items give priority to mastery of processes, understanding of concepts and the ability to function in various situations based on real-life" (Harlen, 2001, p. 82). *Literacy* became the underlying key in the PISA language, as different literacies were assessed in order to judge the extent to which fifteen-year-olds are able to apply what they have learned in school in the outside world. Literacy in PISA means the ability to engage with modern life effectively, and PISA uses real-life contexts to create questions in order to test literacy of three domains (Harlen, 2001).

This will be explored in greater detail in Chapter 4 when we examine the definition of PISA literacy and explore the frameworks used by PISA. For this chapter, it is important to recognize that PISA's assessment of *literacies* rather than *academic content* is a key difference between PISA and IEA studies. Another difference from IEA studies is that PISA attempts to measure the outcome of the whole basic education provided during compulsory education rather than measure the performance of a student at different grade levels (Harlen, 2001).

In other words, PISA is looking to measure what is required for the future, while IEA studies are looking to measure the attainment of the academic school curriculum. PISA sought to identify universally applicable models of schooling to inform best practices specifically related to improving school efficiency (Pereyra, Kotthoff, & Cowen, 2011). As a policy-focused program, PISA was less interested in answering complex research questions and was more interested in presenting data in a simple manner to inform policy in education (Holliday & Holliday, 2003; Morgan & Shahjahan, 2014; Pereyra, Kotthoff, & Cowen, 2011).

PISA and TRENDS Together

PISA and Trends in International Mathematics and Science Studies (TRENDS) continue to be administered on a regular basis every three and four years, respectively. Why the upsurge in international studies?

Some argue that the goal of these studies was reformulated to focus on outcomes that made the studies more feasible. They focused more on evaluating education systems than researching specific questions, and there have been developments in methodology, each of which contributed to the increase in international studies (Gustafsson, 2008).

International agencies are less interested in providing advice, but are now acting as independent agents with educational influence around the world (Mundy & Ghali, 2009; Pereyra, Kotthoff, & Cowen, 2011; Samoff, 2012). In this role, the OECD and IEA are able to rely on their experiences and build upon the authority they already have. Yet in planning and executing future studies, it is also important for both the OECD and IEA not to merely repeat previous successful assessment programs but to embrace the challenge expressed by Husén and Postlethewaite (1996): as education needs and agendas change, international education studies must continue to answer *relevant* questions in education. In their words:

> ... it is the combination of educational research vision and the practical needs of ministry officials that produce worthwhile studies (p. 140).

While Husén and Postlethwaite argued that the future of the IEA depends on the quality of answers it can give, this is certainly pertinent to the OECD as well as any other future international education study endeavor.

PART II

Conducting International Assessments in Mathematics and Science

CHAPTER 3

Who Participates in International Assessments?

Every country in the world has an educational system designed to teach its youth the knowledge, skills, reasoning, and beliefs that are deemed important by that society and its political leadership. The idea behind the group of professors who met in the late 1950s was to develop an international study toward comparing educational systems, especially in terms of academic performance on the common subject matters taught by virtually all countries, e.g. mathematics, language, science, and history.

The previous two chapters of this book traced the history associated with the outgrowth of that original set of meetings. It remained somewhat of a small enterprise for years, run by university professors and serving mostly the academic communities around the subject matters being tested and the emerging area of comparative education. That changed, however, as the Third International Mathematics and Science Study entered the field in the early 1990s. The fact that the US government was involved, coupled with the growing interests of many countries, made international assessments of increasing relevance and importance. Over forty countries participated in various parts of TIMSS-95. The breakup of the Soviet Union also helped ensure the importance of the study because many of the newly independent countries wanted to join the study as they re-examined their own educational systems independently of the former Soviet Union.

This set of events changed the role of international education assessments, as they were no longer viewed as the purview of academics but of governments and the public of participating countries. As the results of TIMSS-95 became available, they were published throughout the world by leading newspapers and news magazines, and broadcast by leading radio and television shows. These studies now belonged to the public and its government. Ministers were held accountable and had to defend poor performance on the part of their countries. Improving performance even became part of political campaign promises for upcoming elections.

The involvement in TIMSS-95 of the governments of major participating countries such as Great Britain, France, Germany, Japan, and the United States, among others, helped to change what had originally been designed as a small-scale mathematics study into a very large enterprise. The US government played a particularly large role in this shift through substantial funding and political support. The influence of the breakup of the Soviet Union as mentioned earlier, together with greater country support and involvement, especially the United States, mainly drove the transformation of international studies in education. As the Organisation for Economic Cooperation and Development (OECD) – a quasi-governmental organization representing thirty-five of the wealthiest nations of the world – began its involvement in the late 1990s, the international comparison of education became a major player in shaping public attitudes about schooling, influencing policy makers around the world and spurring governments to make changes to their curriculum and other policies.

The question for this chapter is to what extent did this interest by governments affect the design of such international assessments? With the success of earlier pre-TIMSS-95 IEA studies came increased attention from governments and the public. As study reports often contained tables with results listed by nations, and many in those nations were not pleased with how their system performed and was represented in the results, explanations were sought in the studies' methodologies, particularly with the way students had been sampled. As a developing field, many of the methodological concerns raised were valid. In addition, the serious time delays in publishing results did not project an aura of competence and authority despite the august reputations of the academics who led the studies. Consequently, participating countries' Ministries of Education (Departments of Education) and related governmental powers became very influential in the planning, execution, and reporting of IEA and OECD studies.

As work began to design the TIMSS-95 study, its funders – mainly the US government – became intimately involved in the details, setting standards for design and execution that the study had to take into account if it wanted to be funded. These standards were the ones the United States used in its own studies such as the National Assessment of Educational Progress (NAEP). This led to much stronger design requirements, especially with respect to sampling, data collection, and reporting requirements. OECD's sponsorship of the first PISA assessment in 2000 reflected similar constraints and requirements. This has both an upside and a downside to it. The great upside is that studies are now run much more efficiently and in keeping with the requirements of survey research. The downside is that more recent studies have become much less innovative than they were when they were viewed as research studies driven by international academic researchers. We will discuss this in more depth in Chapter 11.

Population Definitions

Most international education assessment studies have a simple cross-sectional design. This is true in PISA, and in Pre-TIMSS-95 IEA studies as well as the more recent (post TIMSS-95) Trends in Mathematics and Science Studies (TRENDS). The IEA studies, both pre- and post-TIMSS-95 (TIMSS-95 being a special case which we return to later), typically used a grade-based designation derived from an age-cohort definition. For example, the Second International Science Study (SISS) defined three populations, two of which were specified relative to an age cohort: ten-year-olds and fourteen-year-olds. The third was specified as the final year of secondary schooling. This was done because the age of beginning school varies across countries and there is a lack of standard international grade definitions.[1] As a result, SISS countries were to choose the grade in which the majority of students for each age cohort was to be found. This led to grades 4 or 5 for Population 1, grades 8 or 9 for Population 2, and grade 12 for Population 3.

Table 3.1 gives the population definitions for all of the cross-sectional IEA and OECD international studies focusing on mathematics or science or both. The populations varied in their definitions over the studies as to the focal grades and even the age cohorts from which they were derived, but in all cases, except for the longitudinal part of SIMS, they were all cross-sectional studies (TIMSS-95 being both cross-sectional and cohort longitudinal – see Chapter 9).

The nature of the design has important implications for the analyses and interpretation of the results. Cross-sectional studies are efficiently designed for producing national estimates of performance and other characteristics for the participating countries, thus providing the horse race results with conclusions, such as Singapore has the highest score on the mathematics

---

[1] The International Standards Classification of Education (ISCED), developed in the 1970s and adapted by UNESCO, provides descriptive characteristics for various levels of education, which typically span several years of schooling each.

Table 3.1 *Population definitions of IEA and OECD international mathematics and science studies*

Study	Year	Primary	Lower secondary	Upper secondary
Pilot 12-Country Study	1960		13-year-olds (Seventh and Eighth Grade)	
FIMS	1964		13-year-olds (Eighth Grade)	**End of Secondary** (Twelfth Grade)
FISS (6-Subject Survey)	1970–1971	10-year-olds (Fifth Grade)	14-year-olds (Ninth Grade)	**End of Secondary** (Twelfth Grade)
SIMS	1980–1982		13-year-olds (Eighth Grade)	**End of Secondary** (all students who are in the normally accepted terminal grade of the secondary education system, and who are studying mathematics as a substantial part (approximately 5 hours per week) of their academic program) (Twelfth Grade)
SISS	1983–1984	10-year-olds (Fifth Grade)	14-year-olds (Ninth Grade)	**End of Secondary** (Twelfth Grade)
TIMSS-95	1995	8, 9, 10-year-olds (Third and Fourth Grade)	12, 13, 14-year-olds (Seventh and Eighth Grade)	**End of Secondary** (Twelfth Grade)
TIMSS-R	1999		13 and 14-year-olds (Eighth Grade)	
TIMSS (TRENDS)	2003, 2007, 2011, 2015	9 and 10-year-olds (Fourth Grade)	13 and 14-year-olds (Eighth Grade)	
PISA	2000, 2003, 2006, 2009, 2012, 2015			15-year-olds (Grades 7+, modal grade = 10)

Note. For IEA, populations are normally defined in terms of age, which is then used to identify the particular grades to be sampled, and these can vary by country. The grade levels listed in this table are representative of most countries. The one exception is when the designation "End of secondary" is used instead of an age or grade level. Again, this is because "End of secondary" can be associated with different ages and grade levels in different countries. In PISA, age is used as the selection criterion and only students of that age are sampled, possibly from a range of grades.

assessment. The difficulty arises as researchers try to relate performance within and between countries to other characteristics of the countries and their educational systems. Such relationships can be easily generated but the difficulty occurs in divining their meaning as their study designs typically do not support causal inferences (Schneider, Carnoy, Kilpatrick, Schmidt, & Shavelson, 2007). The problem, however, is that causal inferences are what country government officials, policy people, and academic researchers want in doing these types of analyses. This is a major problem, as those not aware of such limitations or those who choose to ignore them and advocate changes in policy or practice often find out later that such changes did not yield the improvement in achievement predicted by the original analyses.

To go beyond the horse race toward a deeper understanding of why some countries do better than others – to do what the founding fathers were interested in learning and what the comparative education discipline saw as fundamental – demands a different design, one that provides a pre-test so that change over time or gain in performance can be further characterized through statistical modeling. This is the design element that is essential to support causal inference.

Two attempts in that direction have been tried. In SIMS, eight countries (systems) participated in a true longitudinal study in which there was a fall test and then again a test in the spring of the same students (the opposite timeframe was done in the southern hemisphere). SIMS, in the IEA tradition, sampled classrooms and collected data not only on the students but the classroom teacher as well. This put SIMS in the position to (a) be able to link classroom practices, including content coverage (OTL), to gains in student performance or to (b) be able to model performance as a function of the pre-test and instructional characteristics. Both allow for an exploration of learning and growth of eighth-grade students (seventh grade in Japan – the grade where the modal age was thirteen). One entire volume of the three-volume set reporting on SIMS was dedicated to the exploration of the relationships of OTL classroom processes and learning (Burstein, 1993).

Building on the foundation of SIMS, TIMSS-95 wanted to move beyond the horse race and initially considered a longitudinal study toward a better understanding of how opportunity to learn, defined as the implemented curriculum, was related to student learning. In the early design stages of TIMSS-95, Wiley and Wolfe (1992) considered the merits of cross-sectional, longitudinal, and multi-grade designs and made this argument:

> From our perspective, the effects of education are strongly cumulative and all of the learnings that have been attained by a particular age or stage of schooling are a consequence of what has gone on before. Only if longitudinal study designs are used is it conceptually possible to isolate effects that occur within a specified, grade-year-limited time period and even this is a methodologically difficult task. As a consequence, explanations of cross-sectional data on achieved learnings must rest on events that have occurred during the earlier as well as the contemporaneous process of schooling. In our view, adequate scientific evolution of the cross-sectional aspects of IEA studies will occur when better attention is given to these issues (p. 300).

In many ways this was always the goal of IEA studies as Torsten Husén said in the introduction to the First International Mathematics Study:

> ... the main objective of the study is to investigate the "outcomes" of various school systems by relating as many as possible of the relevant input variables (to the extent that they could be assessed) to the output assessed by international test instruments. In the discussions at an early stage in the project, education was considered as a part of a larger social-political-philosophical system. In most countries, rapid changes are occurring, such as revolutionary modifications of the industrial-technical apparatus, patterns of living, geographical and social mobility, attitudes toward the role of science in society, and especially the relative role of humanistic and scientific learning. Any fruitful comparison must take account of how education has responded to changes in the society. One aim of this project is to study how mathematics teaching and learning have been influenced by such developments (1967, p. 30).

As desirable as a longitudinal design was, the judgement was made that it would be too expensive and labor intensive for countries to conduct a full longitudinal study but a cohort-longitudinal study was possible. Following IEA tradition, three populations were defined by choosing the grades associated with an age cohort. The difference between earlier IEA studies and TIMSS-95 was that in TIMSS-95, countries were to choose the two adjacent grades that contained the largest proportion of students at the designated age cohort. This was done to enable cohort-relevant adjustments in the form of a pre-test. Two of the populations were centered on nine-year-olds and thirteen-year-olds. The third population was defined differently: those students enrolled in their final year of secondary school.

Most countries ended up with grades 3 and 4 for Population 1, and 7 and 8 for Population 2. For a small number of countries the grades were 2 and 3 and 6 and 7, respectively. The idea behind this design was the assumption that the two cohorts of students as represented by the adjacent grades were essentially the same in their demographics with the only difference being the additional year of schooling. The upper of the two grades was the focal grade of the study – eighth grade for most countries. Performance in the lower grade provided a pre-test or, in combination with the upper grades, a measure of the gain or growth associated with the focal grade. Since the sampling frame in TIMSS-95 called for the selection of schools in the first stage and a sample of classrooms in the second stage, this made the assumption more reasonable, as the assumption only pertained to the within-school homogeneity of the two cohorts.

As with SIMS, extensive data on OTL was collected from teachers at the upper of the two grades and was related to the gain in performance between the two adjacent grades or as a covariate in a residual gain analysis. The focus of the relational analyses was performance at the upper of the two grades. In a similar way to SIMS, an entire book was written based on these analyses (Schmidt, McKnight, Houang, Wang, Wiley, Cogan, & Wolfe, 2001).

PISA also uses a cohort design but with a focus on fifteen-year-olds. This differs from IEA Studies in that PISA does not move to a focal grade containing the most fifteen-year-olds, but actually draws a true age cohort sample by defining the first stage of the sampling frame as all schools containing fifteen-year-olds. This, for example, results in a variety of samples across the sixty-plus countries that participated in the 2012 study that focused on mathematics. Only two countries (Japan and Iceland) had all of their fifteen-year-olds attending school at the same grade level. For other countries it was not uncommon to have the fifteen-year-old sample drawn from up to five different adjacent grades. Note that PISA's population includes only fifteen-year-olds who are in middle and secondary school, not those who are still in primary school or those who have left school. For many OECD countries, this excludes few fifteen-year-olds, but for some non-OECD countries substantial numbers are excluded.

Tables 3.2 and 3.3 show the results of the sampling for the 2012 PISA. The modal grade for fifteen-year-olds varied from ninth grade in Russia[2] to tenth grade in the United States to eleventh grade in Great Britain and New Zealand. PISA created a variable indicating the modal grade for each country[3] as well the corresponding grade one up and one down from the

---

[2] We use Russia here and throughout the text to refer to the Russian Federation.
[3] We use the word "country," recognizing that in PISA this includes three regions of China (Hong Kong, Macau, and Shanghai), which in the past were often viewed as separate countries, especially in international assessments.

Table 3.2 *Percentage of sampled 15-year-old students at each grade level in the PISA-2012 study: OECD countries*

	All students					
	7th grade	8th grade	9th grade	10th grade	11th grade	12th grade and above
	%	%	%	%	%	%
OECD						
Australia	0.0	0.1	10.8	**70.0***	19.1	0.0
Austria	0.3	5.4	43.3	**51.0***	0.1	0.0
Belgium	0.9	6.4	30.9	**60.8***	1.0	0.0
Canada	0.1	1.1	13.2	**84.6***	1.0	0.1
Chile	1.4	4.1	21.7	**66.1***	6.7	0.0
Czech Republic	0.4	4.5	**51.1***	44.1	0.0	0.0
Denmark	0.1	18.2	**80.6***	1.0	0.0	0.0
Estonia	0.6	22.1	**75.4***	1.9	0.0	0.0
Finland	0.7	14.2	**85.0***	0.0	0.1	0.0
France	0.0	1.9	27.9	**66.6***	3.5	0.1
Germany	0.6	10.0	**51.9***	36.7	0.8	0.0
Greece	0.3	1.2	4.0	**94.5***	0.0	0.0
Hungary	2.8	8.7	**67.8***	20.6	0.0	0.0
Iceland	0.0	0.0	0.0	**100.0***	0.0	0.0
Ireland	0.0	1.9	**60.5***	24.3	13.3	0.0
Israel	0.0	0.3	17.1	**81.7***	0.8	0.0
Italy	0.4	1.7	16.8	**78.5***	2.6	0.0
Japan	0.0	0.0	0.0	**100.0***	0.0	0.0
Korea[1]	0.0	0.0	5.9	**93.8***	0.2	0.0
Luxembourg	0.7	10.2	**50.7***	38.0	0.5	0.0
Mexico	1.1	5.2	30.8	**60.8***	2.1	0.1
Netherlands	0.0	3.6	46.7	**49.2***	0.5	0.0
New Zealand	0.0	0.0	0.1	6.2	**88.3***	5.4
Norway	0.0	0.0	0.4	**99.4***	0.2	0.0
Poland	0.5	4.1	**94.9***	0.5	0.0	0.0
Portugal	2.4	8.2	28.6	**60.5***	0.3	0.0
Slovak Republic	1.7	4.5	39.5	**52.7***	1.6	0.0
Slovenia	0.0	0.3	5.1	**90.7***	3.9	0.0
Spain	0.1	9.8	24.1	**66.0***	0.0	0.0
Sweden	0.0	3.7	**94.0***	2.2	0.0	0.0
Switzerland	0.6	12.9	**60.6***	25.6	0.2	0.0
Turkey	0.5	2.2	27.6	**65.5***	4.0	0.3
United Kingdom	0.0	0.0	0.0	1.3	**95.0***	3.6
United States	0.0	0.3	11.7	**71.2***	16.6	0.2
Number of Countries at Modal Grade	*0*	*0*	*11*	*21*	*2*	*0*

Note. *Indicates modal grade for each country.
Source: PISA 2012 data https://nces.ed.gov/surveys/international/ide/
[1] We use Korea here and throughout the text to refer to the Republic of Korea – South Korea.

Table 3.3 *Percentage of sampled 15-year-old students at each grade level in the PISA-2012 study: non-OECD participants*

	All students					
	7th grade	8th grade	9th grade	10th grade	11th grade	12th grade and above
	%	%	%	%	%	%
Partners						
Albania	0.1	2.2	39.4	58.0*	0.3	0.0
Argentina	2.0	12.0	22.6	59.4*	2.8	1.1
Brazil	0.0	6.9	13.5	34.9	42.0*	2.6
Bulgaria	0.9	4.6	89.5*	4.9	0.0	0.0
Colombia	5.5	12.1	21.5	40.2*	20.7	0.0
Costa Rica	7.4	13.7	39.6*	39.1	0.2	0.0
Croatia	0.0	0.0	79.8*	20.2	0.0	0.0
Cyprus[1]	0.0	0.5	4.5	94.3*	0.7	0.0
Hong Kong-China	1.1	6.5	25.9	65.0*	1.5	0.0
Indonesia	1.9	8.3	37.7	47.7*	3.9	0.6
Jordan	0.1	1.1	6.0	92.9*	0.0	0.0
Kazakhstan	0.2	4.9	67.2*	27.4	0.2	0.1
Latvia	2.1	14.8	80.0*	3.0	0.0	0.0
Liechtenstein	4.9	14.2	66.3*	14.6	0.0	0.0
Lithuania	0.2	6.2	81.2*	12.4	0.0	0.0
Macau-China	5.4	16.4	33.2	44.6*	0.4	0.0
Malaysia	0.0	0.1	4.0	96.0*	0.0	0.0
Montenegro	0.0	0.1	79.5*	20.4	0.0	0.0
Peru	2.7	7.8	18.1	47.7*	23.7	0.0
Qatar	0.9	3.1	13.8	64.8*	17.1	0.3
Romania	0.2	7.4	87.2*	5.1	0.0	0.0
Russia[2]	0.6	8.1	73.8*	17.4	0.1	0.0
Serbia	0.1	1.5	96.7*	1.7	0.0	0.0
Shanghai-China	1.1	4.5	39.6	54.2*	0.6	0.1
Singapore	0.4	2.0	8.0	89.6*	0.1	0.0
Taiwan[3]	0.0	0.2	36.2	63.6*	0.0	0.0
Thailand	0.1	0.3	20.7	76.0*	2.9	0.0
Tunisia	5.0	11.8	20.6	56.7*	5.9	0.0
United Arab Emirates	0.9	2.8	11.3	61.9*	22.2	0.9
Uruguay	6.9	12.2	22.4	57.3*	1.3	0.0
Vietnam	0.4	2.7	8.3	88.6*	0.0	0.0
Number of Countries at Modal Grade	0	0	11	19	1	0

Note. *Indicates modal grade for each country.
Source: PISA 2012 data https://nces.ed.gov/surveys/international/ide/
[1] See Table 3.4 for notes regarding Cyprus
[2] We use Russia here and throughout the text to refer to the Russian Federation.
[3] We use Taiwan here and throughout the text to refer to Chinese Taipei.

mode. Tables 3.2 and 3.3 give the percentage of each country's students falling into each of the six listed grade levels. The modal grade and the related percentage of students found at the grade are highlighted for each country. Given that the students are spread somewhat evenly across the three adjacent grade levels in some countries, the possibility of cohort longitudinal analysis exists in those countries. However, given that the sample of students is not necessarily within the same school, the assumption of homogeneity has to be stronger as the cross-grades homogeneity of demographics is not necessarily exclusively within the same schools. This type of cohort analysis was done in a special paper on OTL for PISA-2012 (Schmidt, Zoido, & Cogan, 2014).

Sampling Design

Both IEA- and OECD-sponsored studies employ a two-stage sampling design in which schools are sampled with probabilities proportional to their size at the first stage. Within countries a layer, or stratum, can be added depending on the educational structure. The strata are often defined geographically, especially with subunits of the government such as regions, provinces, or states. These are typically included only to reduce the variation and are not sampled separately or extensively enough to generate subnational estimates.

In TIMSS (both the TIMSS-95 and the subsequent TRENDS studies) and PISA, subregions in a few countries have been over-sampled so as to generate comparable subregional estimates. In the 2012 PISA study, for example, Italy, Spain, and the United States did this. In the United States, three states opted to participate in this manner – Massachusetts, Connecticut, and Florida. In various TIMSS and TRENDS studies several US states have opted to do this as well.

It is at the second stage in the sampling design where the two studies diverge in their approach. Before discussing the two designs it is important to recognize that the two studies have a different focus, which has implications for the sampling design.

TIMSS-95, as discussed previously, began as a school-based study with the aim of comparing education practices, especially the curriculum, across countries. Even as the results became more publicly and politically relevant, the focus remained on what students learned through their schooling. The assessments were based on what students studied at the different grades across the participating countries.

Since the organizing structure of most schooling around the world is the classroom, all the IEA studies have the classroom as the sampling unit within each school. This led to the collection of data on both students and teachers, thus enabling classroom instruction as informed by the teacher to be linked to student performance. Also, given this focus, specific grade levels, although defined by age cohorts, were chosen for each IEA study, leading to the grade-based population definitions described in the previous section.

PISA on the other hand was more interested in the cumulative knowledge that students acquired across all grades up to and including the grade level they were in as fifteen-year-olds. Given the cumulative focus in terms of the assessment, the particular mathematics or science classroom they were in as fifteen-year-olds was not particularly relevant. Fifteen-year-olds in some countries might not even be taking any mathematics or science classes. In fact, fifteen-year-olds will typically be spread across at least three different grade levels, as illustrated in Tables 3.2 and 3.3, making the classroom even less relevant.

Perhaps an even more salient reason PISA has not focused on classrooms as the sampling unit is the nature of the assessment itself. PISA focuses on mathematics and science *literacy*, not school-based mathematics or science. How students acquire literacy might be in the eyes of some less driven by the school, although some (de Lange, 1996, 2003) have suggested mathematics in schools should be taught almost exclusively in an applied context.

All of this has to some extent led PISA to focus not on the classroom as the sampling unit but on students. In effect, the first stage of sampling draws a sample of all schools containing fifteen-year-old students. Once drawn, the second stage draws a random sample of fifteen-year-old students within the sampled school.

Sampling in a Tracked System

The basic difference between the PISA and IEA sampling designs has to do primarily with the unit that is sampled in the second stage, which in IEA Studies is a classroom in the target grade or grades and in PISA is the individual student without regard to the classroom the student is in. The implications of this with respect to the type of inferences that may be drawn as well as with respect to the measurement of curriculum and opportunity to learn will be addressed in the later chapters. Here we examine the implications of such sampling differences with respect to tracking – a practice found in some countries at the primary or lower-secondary level but found in virtually all countries involved in international assessments around mathematics and science at the upper-secondary level. Tracking at the secondary level typically has to do primarily with student interest and career choices.

Here we focus on content tracking at the primary or lower-secondary level where the purpose is more one of sorting students into different content experiences based on a tested or perceived level of ability. This is not the same as ability grouping in which students cover the same content but at a different pace or with respect to different levels of depth. The key feature in this type of tracking is that the students in different tracks receive content exposure that is designed to be different, such as the coverage of algebra versus advanced arithmetic in different eighth-grade classrooms.

Such tracking is not very common among the countries studied in IEA Studies (PISA has a secondary-level focus) but does exist in one country in particular – the United States. During the 1990s in the United States, tracking was extensively found in the lower-secondary grades (6–9) in mathematics. This was especially true in the seventh and eighth grades, leading to around 25–30 percent of the eighth-grade students being tracked into algebra, while the vast majority of students continued their study of arithmetic, with some placed into what was called pre-algebra – a hybrid course involving arithmetic and some algebra that acted as a substitute class for students of wealthier parents who did not make the cut into a pure algebra course – see Chapter 9 (Cogan, Schmidt & Wiley, 2001; Schmidt, 2009).

TIMSS-95 and the continuing TRENDS studies draw classroom samples in the second stage but they do not stratify the classrooms according to course type (e.g. regular, pre-algebra, Algebra), thus drawing a simple random sample of one, most typically, or two classrooms without explicit regard to course type. Given the goal of generating a national estimate of performance, this would provide an unbiased estimate. However, if the goal is to be able to partition the variance in order to understand where the variation exists so as to develop relevant policy, this approach is inadequate.

Certainly, having only one classroom is totally inadequate as there can be no estimate of between-school classroom variance in performance. Sampling two classrooms does provide an estimate of the between-school variation but that estimate is confounded with tracking, thus limiting the nature of the inference. This is especially important since tracking defined as different content coverage (e.g. algebra vs. arithmetic) would likely account for a sizable amount of the variation in performance depending on the content coverage of the assessment.

Clearly, the policy implications are different depending on whether the majority of the variance is driven by tracking versus differences among classrooms related to student interest, differences in teacher, etc. Studies based on the US TIMSS-95 data where two classrooms were drawn, together with more extensive information on the class structure within each sampled school available from the within-school sampling frame, allowed an estimate of the tracking variance component in performance as differentiated from the classroom variation. This indicated that the tracking variance was 40 percent of the total within-school variance, while classrooms within tracks accounted for only 8 percent – see Chapter 9 (Schmidt, 2009).

At least for the lower-secondary grades, the sampling in the United States and in other countries where tracking is present should include at least two classrooms so as to obtain an estimate of within-school variation but, more importantly, a second stage sampling design that deals with tracking as well when it is present. Although the adoption of the Common Core State Standards in the United States calls for the elimination of tracking in the middle grades, there is some evidence that tracking, although less prevalent now than was the case at the time of TIMSS-95, still exists in the United States and should be addressed in the sampling approach for future TRENDS studies. Unfortunately, the TRENDS studies have moved away from the TIMSS-95 cohort-longitudinal design and, for example, the recent 2015 version sampled only one classroom per school.

Country Participation

Conducting IEA or OECD studies is expensive for the participating countries and, as a result, participating countries are self-selected, i.e. they are never a random sample of countries. For political reasons or reasons of cost, some major countries such as India and China, together representing over 40 percent of the world's population, have not participated or have had only certain cities or regions participate, such as Shanghai, Hong Kong, and Macau in China. In this section, we provide a summary on who has participated and in which study.

Table 3.4 Participation in the PISA and IEA mathematics and science studies

Country	Pilot	FIMS	FISS	SIMS	SISS	TIMSS1995	TIMSS1999	TIMSS2003	TIMSS2007	TIMSS2011	TIMSS2015	TIMSS2019♦	PISA2000	PISA2003	PISA2006	PISA2009	PISA2012	PISA2015	PISA2018♦	Total
Albania												●	●			●	●	●	●	6
Algeria															●				●	3
Argentina						●		●	●	●	●	●	●		●	●	●		●	9
Armenia								●	●	●	●	●							○	5
Australia		●	●	●	●	●		●	●	●	●	●	●	●	●	●	●	●	●	17
Austria		●		○		●	●		●	○		●	●	●	●	●	●	●	●	11
Azerbaijan										○	○	○				●			○	5
Bahrain								●	●	●	●	●								5
Belarus																			●	1
Belgium	●		●		●	●	●	●	●	●	●	●	●	●	●	●	●	●	●	17
Bosnia and Herzegovina									●							●	●			3
Botswana								●		●	●	●								4
Brazil									●				●	●	●	●	●	●	●	7
Brunei Darussalam												●								1
Bulgaria						●	●		●		●	●		●	●	●	●	●	●	12
Canada				●		●	●	○	○	○	●	●	●	●	●	●	●	●	●	16
Chile						●		●		●	●	●			●	●	●	●	●	12
China					○											○	○	○		4
Colombia						●			●						●	●	●		●	7
Costa Rica																●*	●	●	●	4
Croatia												●			●	●	●	●	●	8
Cyprus¹						●	●		●	●	●	●							●	8

Country	13	12	2	4	1	6	16	14	7	15	4	8	1	16	17	8	2	13	8	11	17	16	18	11	8	15	3	5	2	11
Czech Rep.	•	•	•				•	•	•	•		•		•	•	•		•		•	•	•	•	•	•	•				•
Denmark	•	•	•				•	•	•	•		•		•	•	•		•		•	•	•	•	•	•	•				•
Dominican Rep.	•	•					•	•	•			•			•	•		•		•	•	•	•	•		•				•
Egypt	•	•					•	•	•	•		•		•	•	•	○	•		•	•	•	•	•	•	•			•	•
El Salvador	•	•					•	•	•	•				•	•	•		•		•	•	•	•	•		•			•	•
Estonia	•	•					•	•		•				•		•		•		•		•		•	•	•				•
Finland	•	•					•	•		•				•	•	•		•		•		•		•	•	•				•
France	•	•		•			•	•	•	•				•	•					•	•	•	•	•	•	•				
Georgia	•	•		•			•	•	•	•				•	•					•	•	•	•	•	•	•				
Ghana	•	•						•	•	•		•		•	•					•	•	•	•	•	•	•				•
Greece	•	•		•	•			•	•	•				•	•					•	•	•	•	•		•				•
Honduras					•				•					•	•						•	•	•	•		•				•
Hong Kong	•							•						•	•						•	•	•	•		•				•
Hungary	•	•						•		•		•		•							•	•	•	•		•				•
Iceland															•						•	•	•			•				
India								•	•					•	•						•	•								
Indonesia							•	•	○						•		○			•	•									
Iran							•	•	○											•	•									
Ireland							•	•	○												•									

Table 3.4 (cont.)

Country	Pilot	FIMS	FISS	SIMS	SISS	TIMSS1995	TIMSS1999	TIMSS2003	TIMSS2007	TIMSS2011	TIMSS2015	TIMSS2019♦	PISA2000	PISA2003	PISA2006	PISA2009	PISA2012	PISA2015	PISA2018♦	Total
Lebanon								•	•	•	•	•						•	•	7
Liechtenstein													•	•	•	•	•	•	•	6
Lithuania						•	•	•	•	•	•	•			•	•	•	•	•	12
Luxembourg				•									•	•	•	•	•	•	•	8
Macao														•	•	•	•	•	•	6
Macedonia							•	•		•		•	•					•	•	7
Malaysia							•	•	•	•	•	•				*•	•	•	•	10
Malta							•			•	•	•				*•		•	•	7
Mauritius																*•				1
Mexico						•							•	•	•	•	•	•	•	8
Moldova							•	•	•							*•		•		5
Mongolia									•											1
Montenegro												•			•	•	•	•	•	6
Morocco						•	•	•	•	•	•	•								7
Netherlands		•	•	•	•	•	•	•	•	•	•	•	•	•	•	•	•	•	•	17
New Zealand			•	•	•	•	•	•	•	•	•	•	•	•	•	•	•	•	•	16
Nigeria			•		•															2
Norway						•	•	•	•	•	•	•	•	•	•	•	•	•	•	14
Oman									•	•	•	•								4
Pakistan												•								1
Palestinian Nat'l Auth.								•	•	•										3
Panama																•			•	2

60

Country	Total
Papua New Guinea	1
Peru	5
Philippines	6
Poland	12
Portugal	11
Qatar	9
Romania	12
Russia	14
Saudi Arabia	6
Serbia	9
Serbia and Montenegro	1
Singapore	12
Slovak Rep.	13
Slovenia	12
South Africa	5
Spain	13
Swaziland	1
Sweden	18
Switzerland	9
Syria	3
Taiwan	11
Thailand	15
Trinidad and Tobago	3
Tunisia	9
Turkey	11
Ukraine	3
United Arab Emirates	8
United Kingdom[2]	19
United States	19
Uruguay	6
Venezuela	1

Table 3.4 (cont.)

Country	Pilot	FIMS	FISS	SIMS	SISS	TIMSS1995	TIMSS1999	TIMSS2003	TIMSS2007	TIMSS2011	TIMSS2015	◆TIMSS2019	PISA2000	PISA2003	PISA2006	PISA2009	PISA2012	PISA2015	◆PISA2018	Total
Vietnam	●																●	●	●	3
Yemen								●	●	●										3
Yugoslavia																				1
Zimbabwe					●															1

Note. The names of the countries on this table are shortened where possible to facilitate easy location of the countries.
● = A country participated and reported as a whole country in at least one available component of the assessment in the specified year.
○ = Part of the country, such as a region or a city or a group of regions or cities, participated and reported in at least one available component of the assessment in the specified year and the country did not report as a whole country.
*for PISA-2009 indicates that the country participated in 2010.
◆for PISA-2018 and TIMSS-2019 the countries indicated are scheduled to participate.
1. For Cyprus: In PISA, Cyprus is listed as a participating country. OECD provided the following footnotes:
"Footnote by Turkey: The information in this document with reference to 'Cyprus' relates to the southern part of the Island. There is no single authority representing both Turkish and Greek Cypriot people on the Island. Turkey recognises the Turkish Republic of Northern Cyprus (TRNC). Until a lasting and equitable solution is found within the context of the United Nations, Turkey shall preserve its position concerning the 'Cyprus issue'."
"Footnote by all the European Union Member States of the OECD and the European Union: The Republic of Cyprus is recognised by all members of the United Nations with the exception of Turkey. The information in this document relates to the area under the effective control of the Government of the Republic of Cyprus."
(OECD 2018, found www.oecd.org/pisa/aboutpisa/cyprus-pisa.htm)
2. For the United Kingdom: when participating as a whole country, the participants may include any combination of one or more of the following: England, Northern Ireland, Scotland, and/or Wales. When participating as a part of the country, the participants may include any combination of one or more of the following: England, Northern Ireland, Scotland, and/or Wales, and report as individual entities. This follows the OECD reporting, where the United Kingdom participated in, and reported scores as, one entity.

In the first of the IEA studies prior to SIMS and SISS, the number of participating countries was relatively small, ranging from twelve to nineteen countries or regions of countries. SIMS and SISS were also relatively small (around twenty-two to twenty-three), but beginning with the TIMSS-95, the number of participating countries greatly expanded to more than forty. The most recent 2015 studies of both TRENDS and PISA each had more than fifty and seventy participants, respectively.

The First International Mathematics Study included countries such as Australia, Belgium, England, Germany, France, Sweden, Japan, the Netherlands, and the United States, while the Second International Mathematics Study added countries such as Hong Kong and New Zealand. The TIMSS-95 added most of the remaining European countries as well as Korea[4] and Canada. Much the same pattern of country participation was evident for the science studies. See Table 3.4 for a complete listing of countries participating in the seventeen mathematics and science studies as well as those scheduled to participate in the PISA-2018 and TIMSS-2019 studies. As noted in the table, parts of countries, such as several Canadian provinces, and not the whole country were participants in some studies (and the results were reported by subregions).

Over one hundred countries have participated in at least one of the seventeen IEA or OECD studies (or are scheduled to participate in the next PISA or TRENDS study). Only the United States and the United Kingdom (or some parts of the United Kingdom, especially England) have participated in all seventeen studies. Looking back at the country identification of the founding fathers, this is not surprising. Japan has participated in all studies, with the only exception being the original Pilot Study. Other countries who were involved in a substantial majority of the studies included: Australia, New Zealand, Italy, the Netherlands, Norway, Hungary, Korea, and Thailand.

Types of Data Collection

Each of the IEA and OECD studies collected substantial amounts of data in addition to the actual student assessments, including student, teacher, principal/headmaster, school, and parent questionnaires. They measured such broad categories as attitudes, values, experiences, demographics, aspirations, and types of support to name a few. These are often specific to individual studies and reflect particular study interests.

---

[4] We use Korea here and throughout the text to refer to the Republic of Korea – South Korea.

However, all of these studies have three things in common – measures of student socioeconomic status, performance, and opportunity to learn (the latter became a part of PISA in 2012). In the next chapter we focus on the assessment part of the studies, and subsequent chapters are dedicated to the OTL and SES components.

CHAPTER 4

What Students Know
From Items to Total Scaled Scores

From the beginning with the original IEA Pilot Study, the focus has been, as Torsten Husén stated, on "developing a methodology for quantitative comparative studies" with the emphasis being on the quantitative nature of those studies (1979, p. 373). Comparative education studies up to around 1960 had pretty much been historical or qualitative in nature and not based on large-scale survey research methodology, which itself was coming into prominence in the broader social sciences at about the same time. One good example illuminating the growing attention to the large-scale surveys was the US study of schools done by sociologist James Coleman in 1966 (Coleman et al., 1966).

The emphasis on the quantitative was of paramount importance to the founding fathers:

> The bold idea of conducting a study of cognitive development in children belonging to different national systems of education was first brought up at a meeting of educational researchers from a dozen countries at the UNESCO Institute for Education in Hamburg in 1958.[12] The year before, that institute had hosted an international meeting of educational psychologists – most of them psychometricians – on problems of evaluation. This was a field in Europe to which little thought had been devoted at that time. In the United States, through Ralph Tyler's pioneering research, evaluation had long been an area in which educators took great interest.[13]
>
> At a previous meeting in 1957 it was realized how little empirical evidence was available to substantiate the sweeping judgments that were commonplace about the relative merits and failings of various national systems of education. Concerns about the quality of secondary education in general – science education in particular – had begun to be aired in the United States by Admiral Rickover and the history professor Arthur Bestor.[14] American schools were under attack, accused of a lack of intellectual rigor and standards. Similar concerns had begun to crop up in other countries where secondary education was in the process of becoming universal (Husén, 1979, p. 373).

The influence of Tyler, Thorndike, Coleman, Plowden, and others led to the belief that valid and reliable international measures of cognitive development could be created. Consequently, the Pilot Study of 1960 included measures of reading comprehension, arithmetic, science, and geography. Later studies have included other areas such as: English as a Second Language, French as a Second Language, and Literature, but after the 1970s the subject matters most often assessed in both the IEA and OECD studies have focused on mathematics, science, and reading. Nonetheless, both the IEA and OECD have carried out other international studies related to education such as those listed here:

IEA
- Reading (Progress in International Reading Literacy Study - PIRLS)
- Advanced mathematics and physics for students in their final year of secondary school
- Computer and information technology in education (ICILS)
- Early Childhood Education Study (ECES)
- International Civic and Citizenship Study (ICCS)
- International Study of Teacher Preparation in Mathematics for primary and lower-secondary teachers in eighteen countries (Teacher Education and Development Study in Mathematics - TEDS-M)

OECD
- Teaching and Learning International Survey (TALIS)
- PISA for Development
- PISA-based Test for Schools
- Survey of Adult Skills (PIAAC)
- International Early Learning and Child Well-being Study (IELS)

The focus in this book is on the K–12 assessments of mathematics and science only. To include PISA reading, for example, would require expertise and experience not present among us – the authors. Additionally, covering both mathematics and science in detail would be too large a task and one that would produce an encyclopedia-like book, not an overview of international assessments. To achieve a compromise, both areas are discussed but greater elaboration is reserved for mathematics, it being next to reading as the second fundamental literacy and, in general, of more interest to countries.

In what follows both in this chapter and in the next four chapters, we focus on three of the major defining elements of schooling and common components of international assessments of mathematics and science: the assessments themselves; the curriculum or content coverage measures which we characterize as opportunity to learn (OTL); and measures of the home and family background demographics of the students (SES). The IEA and PISA studies both have other aspects of schooling that the questionnaires address such as attitudes, university and career expectations, and values of the students together with characteristics of the teacher, including their training and instructional practices, as well as school questionnaires including school structures and policies. We do not address these in this book as their occurrence, concentration, and the methodology used to measure them vary for different studies. These contextual issues provide an important backdrop yet our focus is on the unfolding center stage drama of schooling: students' background and what they learn from the learning opportunities they have had. Measures of these essentials should be (and mostly are now given PISA's inclusion of OTL in 2012) included in every international assessment of K–12 student learning.

As we address each of the big three areas, we look historically as to how they were conceptualized and measured over the various mathematics and science studies up to the most current PISA and IEA studies. We do not provide detailed historical development of each of the seventeen studies but enough to show the evolution of the methodology to the present.

Chapters 1–3 gave a brief overview of the studies to provide a general background but the next four chapters focus not on the studies themselves but on the methodologies employed in the various studies as they addressed the measurement of schooling (OTL), home/family background (SES), and student performance. The final three chapters (9–11) examine what we have learned both substantively and methodologically across the seventeen studies (consisting of the IEA studies of mathematics and science and the PISA mathematics and science literacy studies).

To accomplish this we focus on FIMS, SIMS, TIMSS-95, and the five TRENDS to represent the eight mathematics international assessments to date and FISS (the science part of the Six Subject Survey), SISS, TIMSS-95, and TRENDS to represent the eight international science assessments. The Pilot Study and the five other areas of the Six Subject Survey were only included in Chapter 1 for historical completeness. For PISA, we have the 2003 and 2012 mathematics-focused studies, and the 2006 and 2015 science-focused studies. We exclude the 2000 and 2009 reading-focused studies.

Achievement vs. Literacy

Both IEA and PISA test mathematics and science but what these assessments are designed to measure is not the same. IEA studies focus on school or academic learning as it relates to the curriculum. Each is a study of schooling focusing on understanding the content coverage (OTL) and nature of the instruction provided by the teacher and its relationship to performance. This is why the sampling, described in Chapter 3, focuses on the classroom and not the individual student. It is why all IEA studies measure the curriculum in some fashion, since performance on the tests is expected to reflect those curricular experiences.

PISA, on the other hand, is more concerned with what students can do with the knowledge acquired through their schooling. It does not develop its tests by examining the curriculum of participating countries but rather by first defining a model of literacy, i.e. the ability to use the science or mathematics learned in school to address and solve real-world problems. The PISA-2012 Model of Mathematical Literacy made explicit the process by which real-world problems may be solved (see Figure 4.1). This model reflects the 2012 definition of mathematical literacy:

> Mathematical literacy is an individual's capacity to formulate, employ, and interpret mathematics in a variety of contexts. It includes reasoning mathematically and using mathematical concepts, procedures, facts and tools to describe, explain and predict phenomena. It assists individuals to recognize the role that mathematics plays in the world and to make the well-founded judgments and decisions needed by constructive, engaged and reflective citizens (OECD, 2013a, p. 25).

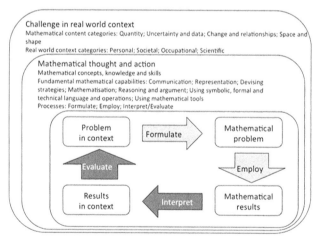

Figure 4.1 A Model of Mathematical Literacy in practice.
Note. Republished with permission of OECD, *PISA 2012 Assessment and Analytical Framework: Mathematics, Reading, Science, Problem Solving and Financial Literacy.* OECD, 2013a; permission conveyed through Copyright Clearance Center, Inc.

Real-world problems are characterized in PISA by the context in which the problem exists (societal, personal, occupational, etc.) and the nature of the mathematics that underlies the problem. The detail of the model is described as follows:

> The outer-most box in figure 1.1 [Figure 4.1 in this text] shows that mathematical literacy takes place in the context of a challenge or problem that arises in the real world. In this framework, these challenges are characterised in two ways. The context categories, which will be described in detail later in this document, identify the areas of life from which the problem arises. The context may be of a *personal* nature, involving problems or challenges that might confront an individual or one's family or peer group. The problem might instead be set in a *societal* context (focusing on one's community – whether it be local, national, or global), an *occupational* context (centered on the world of work), or a *scientific* context (relating to the application of mathematics to the natural and technological world). A problem is also characterised by the nature of the mathematical phenomenon that underlies the challenge. The four mathematical content categories identify broad classes of phenomena that mathematics has been created to analyse. These mathematical content categories (*quantity, uncertainty and data, change and relationships,* and *space and shape*) are also identified in the outer-most box of figure 1.1 [Figure 4.1 in this text] (OECD, 2013a, pp. 25–26).

Science literacy is similarly defined in terms of addressing real-world problems but due to the nature of the science disciplines it is described differently than mathematics. The following definition was developed in the 2015 PISA assessment, which focused on science.

> Scientific Literacy is the ability to engage with science-related issues, and with the ideas of science as a reflective citizen.
>
> A scientifically literate person, therefore, is willing to engage in reasoned discourse about science and technology which requires the competencies to:
>
> 1. Explain phenomena scientifically:
>   - Recognise, offer and evaluate explanations for a range of natural and technological phenomena.
> 2. Evaluate and design scientific enquiry:
>   - Describe and appraise scientific investigations and propose ways of addressing questions scientifically.
> 3. Interpret data and evidence scientifically:
>   - Analyse and evaluate data, claims and arguments in a variety of representations and draw appropriate scientific conclusions (OECD, 2015, p. 7).

Content Frameworks

A hallmark of all IEA international assessments is the presence of a framework or, as it is often called, a grid, which most typically is conceived of as a matrix listing the topics or content as one of the dimensions. A second dimension is meant to characterize the type of cognitive behavior or, as defined in the original TIMSS-95, the nature of the performance expectation required of the student. Both characterizations of the second dimension are intended to describe what students are expected to do with the content, know, apply, reason with, etc. This was the case in both the First International Mathematics Study (FIMS) and the Second International Science Study (SISS) but in the former a third dimension representing the different types of applications of mathematics was added. This dimension, however, was never actually utilized in the study.

The second studies (SIMS and SISS) and the third study (TIMSS-95) also developed two-dimensional grids with similar second dimensions. TIMSS-95 also had a third dimension, which was termed "perspectives" on mathematics and science but it, like in FIMS, was never really used to any great extent in the study. The five TRENDS studies that followed TIMSS-95 were also based on similar two-dimensional grids. In each study, the grid or framework was intricately tied to the test blueprint from which items were to be developed.

The development of the frameworks for the first two rounds of IEA studies involving mathematics and science was based on an informal survey of the participating countries using whatever they could such as textbooks, tests, and expert judgement, asking them what topics were covered in their country at the populations being tested. This led to a draft version of a framework for each population. The draft was sent to the participating countries, who were asked to indicate in some form if the list of topics was appropriate to that country's intended curriculum for the tested populations. For example, in the Second International Mathematics Study (SIMS) the national centers were asked to indicate, using a three-point scale, the degree of emphasis each topic should have: very important internationally; important to most countries; and important to some countries. Following the review, further revisions ensued to produce the final version of the framework.

The Third International Mathematics and Science Study (TIMSS-95) approached the development of the framework differently, expanding on the idea of the earlier studies' use of a two-dimensional grid for each tested population. The TIMSS-95 frameworks were designed not to be

What Students Know

population specific but to represent mathematics and science content across all grades K–12 (Robitaille, Schmidt, Raizen, McKnight, Britton, & Nicol, 1993; Schmidt, McKnight, Valverde, Houang, & Wiley, 1997; Schmidt, Raizen, Britton, Bianchi, & Wolfe, 1997). They were also not primarily designed to be the test blueprints.

One of the first tasks to be undertaken in TIMSS-95 was to convene a group of mathematicians, mathematics educators, scientists, and science educators to work on the development of the draft content frameworks. Completed drafts were then reviewed by all participating countries (this phase of the work included around fifty countries). The national centers were asked to identify five topics in their K–12 curriculum that were not represented in the two frameworks. If there were any such topics, they were added in the next draft. In effect, the resulting frameworks were designed to represent the union of all topics intended for all the grades K–12 over all participating countries and not only what was in common to all countries – in other words not the intersection of topics covered worldwide.

The two frameworks were developed not only to guide the development of the test blueprints but also to provide the means by which all data collection activities, including the OTL measures, the questionnaire items, and the curriculum analysis procedures, would have a common metric having been mapped onto the same mathematics and science framework. This is best summarized as follows:

> For the purposes of TIMSS, curriculum consists of the concepts, processes, and attitudes of school mathematics and science that are intended for, implemented in, or attained during students' school experiences. Any piece of curriculum so conceived-whether intended, implemented, or attained, whether a test item, a paragraph in an "official" curriculum guide, or a block of material in a student textbook-may be characterized in terms of three parameters: subject-matter content, performance expectations, and perspectives or context (Robitaille et al., 1993, p. 43).

The two frameworks are further discussed in Chapter 7 relative to opportunity to learn (OTL) and the TIMSS-95 curriculum analysis. The actual frameworks are also listed in Appendix A. The blueprints for test development for each of the TIMSS-95 three populations were weighted subsets of each of the three dimensions of the framework.

This TIMSS-95 framework process was identical for mathematics and science. A group of scientists worked along the same timeline as the mathematics group did in order to develop the corresponding science framework, which was composed of the same three dimensions. It also served the same function of providing for study coordination across all data collections as well as providing the test blueprint.

The PISA mathematics framework has three dimensions: mathematics content knowledge; contexts; and mathematical processes (underlying mathematical capabilities). The mathematics content areas represent the traditional areas of the mathematics curriculum – algebra, geometry, data, and number. However, because the PISA assessment is about literacy and the use of such mathematics to solve real-world problems, these four areas are defined more in terms of broad areas of phenomena that can be aggregated into areas of mathematical thought. Put another way:

> Mathematical structures have been developed over time as a means to understand and interpret natural and social phenomena. In schools, the mathematics curriculum is typically organised around content strands (e.g. number, algebra and geometry) and detailed topic lists that reflect historically well-established branches of mathematics and that help in defining a structured curriculum. However, outside the mathematics classroom, a challenge or situation that arises is usually not accompanied by a set of rules and prescriptions that shows how the challenge can be met. Rather it typically requires some creative thought in seeing the possibilities of bringing mathematics to bear on the situation and in formulating it mathematically. Often a situation can be addressed in different ways drawing on different mathematical concepts, procedures, facts or tools.
>
> Since the goal of PISA is to assess mathematical literacy, an organizational structure for mathematical content knowledge is proposed based on the mathematical phenomena that underlie broad classes of problems and which have motivated the development of specific mathematical concepts and procedures. For example, mathematical phenomena such as uncertainty and change underlie many commonly occurring situations, and mathematical strategies and tools have been developed to analyse such situations. Such an organisation for content is not new, as exemplified by two well-known publications: *On the Shoulders of Giants: New Approaches to Numeracy* (Steen, 1990) and *Mathematics: The Science of Patterns* (Devlin, 1994) (OECD, 2013a, p.31).

As a result the mathematics content, dimension is defined as having four categories: change and relationship, space and shape, quantity, and uncertainty and data. The literacy orientation of the items to be developed is required to have not only a connection to one of these four areas of mathematics but also to a real-world context out of which the problem arises. PISA identifies four such context areas: personal, occupational, societal, and scientific.

To understand the third dimension of the framework, we return to the definition of mathematical literacy as represented in Figure 4.1. Three main processes are defined relative to solving a problem: formulation of the problem in mathematical terms; solving the formulated mathematics problem (employing); and interpreting the mathematical solution arrived at in terms of the original problem. These three dimensions are reported as PISA sub-scores. The test blueprint suggests a distribution of score points of 25 percent formulating, 50 percent employing, and approximately 25 percent interpreting (OECD, 2013a, page 38, table 1.1). Score points are used instead of the number of items because PISA items are scored at different levels of performance (not just dichotomously as right/wrong), and as such emphasis was measured by the number of score points associated with an area of the assessment.

At one level of abstraction, the PISA framework is quite similar to the original Third International Mathematics and Science Study in that it is more broadly and conceptually defined, rather than focused narrowly on the world of fifteen-year-olds. In another way, two of the dimensions are very similar to all IEA studies in mathematics and science: one dimension is about content but more phenomenally rather than disciplinarily defined; the other about cognitive processes. Because PISA focuses on real-world problems it necessitates the addition of the third dimension.

Sampling the Domain: Breadth vs. Depth in Terms of Domain Coverage

The previous sections describe how the intended assessment purpose and focus and the attendant frameworks come to define what it is that is to be measured in international assessments. They all develop a framework, most often with two or three dimensions, which, when combined, explicitly define the content domain of interest. The question of this section is how do they sample the domain? The answer to this question is intertwined with the limiting factor found in every participating country – how much student time is available (usually defined by the country) for the assessment? Put simply, how long can the test be?

The answer, in general, seems to be about one to two hours for international assessments. In PISA this includes all three tests (mathematics, science, and reading), while in IEA TIMSS-95 and TRENDS Studies it includes both the mathematics and science tests. The inherent problem that arises is that testing time and the number of items are positively related but time itself has imposed limits. Subject matter experts typically suggest that the study needs a large number of items in order to adequately represent the domain as defined by the content framework.

Second, psychometricians want a substantial number of items for the measurement of any subpart of the domain for which a subtest will be developed. The two combined imply that the number of test items needed to estimate student performance using all cells in the domain would take more testing time than is available. Some of the frameworks used in the earlier IEA studies had upward of 100 topics defining the content domain alone, along with an additional four or five categories for the cognitive process dimension, resulting in up to a 400-cell content domain. We look at the three mathematics studies to see how this has been handled.

To begin an examination of this issue we look briefly at the original Pilot Study where the 120-item omnibus test had not only arithmetic items but also items assessing geography, reading comprehension, and science, leaving a relatively small number of items dedicated to mathematics. The first "full" study was the First International Mathematics Study (FIMS), in which four defined student populations were to be tested by all participating countries (thirteen-year-olds in the modal grade – Level 1; all thirteen-year-olds in any grade – Level 1b; end of secondary–mathematics – Level 3a; and end of secondary–non-mathematics – Level 3b). Also included in the study was an optional population – fifteen- to sixteen-year olds – Level 2. This involved the development of four different test forms – one for each student population. The tests were designed to be completed in around one hour. Table 4.1 shows how the mathematics content was distributed across the four populations. A single test form was developed for each population, resulting in roughly sixty to seventy-item tests.

In FIMS the final tests for the different populations had around seventy items, which to the researchers seemed too few to adequately portray country differences in mathematics achievement. The limitation on the number of items was related to the amount of time available for testing individual students. This problem arose because all students in a given population were to take the same test.

By the time of SIMS, some twenty years post-FIMS, test methodology had advanced, allowing for more items to be administered in the aggregate so as to get a richer description of student performance at the national level. In SIMS the notion of matrix sampling and rotated forms was employed to address the problem. No student in SIMS took all the available items but only a subset of them. The forms were then randomly distributed across the sampled students. As a result, at the country level and across all students the estimate of achievement could be based on a much larger number of items allowing for a more complete and expansive coverage of the mathematics domain.

Table 4.1 *Number of items addressing each mathematical content area for different student levels (populations) in FIMS*

Topic	Level 1	Level 2	Level 3a	Level 3b
Basic arithmetic	13	3		3
Advanced arithmetic	18	7	3	9
Elementary algebra	12	6	1	5
Intermediate algebra	4	16	19	13
Euclidian geometry	13	17	5	13
Analytic geometry	1	4	8	5
Sets	4	3	4	4
Trigonometric and circular functions		1	3	3
Analysis			8	1
Calculus			9	
Probability			1	1
Logic		2	8	1
Affine Geometry	3			

Note. From Husén, T., *International Study of Achievement in Mathematics: Comparison of Twelve Countries* (Vol 1.), 1967, p. 105. Copyright 1967, International Project for the Evaluation of Educational Achievement, Hamburg Almqvist & Wiksell, Gebers Förlag AB, Stockholm. Publisher: John Wiley & Sons Inc.

The SIMS design created five test forms, with a core form that was to be taken by all students. Students were randomly assigned to two forms – the core and one rotated form. Interestingly, this resulted in around a seventy-item test like FIMS for most students, but at the country level estimating the average achievement level of thirteen-year-olds was based on 176 items. This not only spanned the domain more completely but also provided more items and hence greater reliability in the estimation of sub-scores. For the subset of six of the eight countries that participated in the longitudinal version of SIMS, the same test forms and items were used in the pre-test as well as in the post-test. In the other two countries the composition of the pre-test included only a subset of the items.

TIMSS-95 and the five TRENDS studies thereafter followed in the footsteps of SIMS by using rotated forms. TIMSS-95, for example, expanded the number of forms for each population to "eight booklets of approximately equal difficulty and equivalent content coverage" (Martin, & Kelly, 1996, pp. 2–16). A core of items appeared in all booklets. The development also included link items across Populations 1 and 2 and between Populations 2 and 3. SIMS items were also included in the Population 3 TIMSS-95 Mathematics test booklets. The resulting booklets were based on a total of 102 and 151 available items, respectively, for Populations 1 and 2 (see Table 3.1 for the population definitions). The numbers for science were 97 and 135 items.

PISA uses a similar rotated booklet design:

> The paper-based instruments for the PISA 2012 survey contain a total of 270 minutes of mathematics material. The material is arranged in nine clusters of items, with each cluster representing 30 minutes of testing time. Of this total, three clusters (representing 90 minutes of test time) comprise link material used in previous PISA surveys, four "standard" clusters (representing 120 minutes of test time) comprise new material having a wide range of difficulty, and two "easy" clusters (representing 60 minutes of test time) are devoted to material with a lower level of difficulty.
>
> Each participating country used seven of the clusters: the three clusters of link material, two of the new 'standard' clusters allows for better targeting of the assessment for each of the participating countries; however, the items are scaled in such a way that a country's score will not be affected if it chooses to administer either the "easy" or additional "standard" clusters. The item clusters are placed in test booklets according to a rotated test design, with each form containing four clusters of material from the mathematics, reading and science domains. Each student does one form, representing a total testing time of 120 minutes (OECD, 2013a, p. 39).

The Items

The development of test items is a laborious and time-consuming task requiring piloting and usually a field trial, where the conceptual design of the forms described in the previous section are merged with the items that have survived the pilot testing. The Pilot Study, which began the IEA's international assessment work, found itself with, "no time for a laborious, time-consuming exercise of test development, those in the group who were experts in test development drew upon items already available, most of them from England and the United States" (Husén, 1979).

The first full-blown IEA study, FIMS, found itself in a similar situation and ended up relying heavily on two US sources: the Educational Testing Service (ETS) and the University of Chicago's Examiner's Office. Ultimately, they had an item pool of over 600 items. Similar experiences occurred in the other early IEA studies, except by the time of the second studies in mathematics (SIMS) and science (SISS) the sources had become more democratically dispersed – the National Centers from the participating countries. Once again a major source was the United States, which included the released items from the National Assessment of Educational

Progress (NAEP). Items from FIMS were also included. The project itself developed items as well. For the two populations studied in SIMS, the initial item pools were 480 and 400. SISS followed a similar procedure that relied on countries to provide items.

The final selection of items to be included in the actual assessments was constrained to some degree after the first mathematics and science studies as researchers wanted to be able to establish trends over time. As such, the second studies (SIMS and SISS) had to include a substantial number of items in their final selection that had been previously administered in FIMS and FISS.

For the SIMS cross-sectional part of the study this led to 176 items (180 for the longitudinal part of the study) for the younger population and 136 for the older population, which were then allocated to the various forms. For SISS the final pool of items for allocation to the different forms for the three populations were 56, 70, and 150, respectively (Rosier & Keeves, 1991; Travers & Westbury, 1989).

With TIMSS-95 several changes occurred in the IEA traditional approach to item development. Prior to this, the items were always multiple choice with five alternatives. TIMSS-95 introduced two new types of assessment items: open-ended with both extended and short responses and performance assessment tasks, the latter being included only in a special sub-study (see Chapter 2).

In addition, the development process itself was more extensive and expansive. The initial development of the item pool was done very similarly to past IEA studies. Items from SIMS and SISS were included as well as items provided by the national centers. An item pilot yielded 137 and 279 acceptable mathematics items for each of Populations 1 and 2, but much criticism of the pool followed. This led the IEA to contract with ETS to develop new items as well as with the Australian Council for Educational Research (ACER) to produce additional Population 3 items. These were further piloted (in an already scheduled field trial) and then further reviewed by international committees, who felt yet more items were needed in certain areas. This led to yet another round of contracts to produce supplementary items. Because of the tight timeline, no further piloting was possible but many of the items submitted by the contractors had already been piloted in other studies.

This led to the longest item development process in the IEA's history, supported by external contractors who hired professional item writers. This was made possible by the extensive involvement of the US government, including the availability of the necessary funding. This process moved a long way from what was done in the past: involving professional item writers to complete the needed item pool. Both mathematics and science item developments were done in the same fashion. As part of the process, link items were developed and included to link Populations: 1 and 2; 2 and 3; 3 and SIMS Population A; and 3 (advanced mathematics) and SIMS Population B.

One of the least well-known aspects of the TIMSS-95 study was the performance assessment study in both mathematics and science done by a smaller number of countries (see Chapter 2). Thirteen tasks (six science, five mathematics, and two combined) were developed consistent with the TIMSS-95 content framework:

> The performance assessment tasks required students to engage in an experimental procedure or manipulation of equipment during which they responded to a number of task-related questions (hereafter referred to as "items"). Each task generally began with a statement defining a central problem or investigation, such as "Investigate what effect different temperatures have on the speed with which the tablet dissolves" (Martin & Kelly, 1996, p. 6–2).

The second major change in TIMSS-95 that impacted the item development process was that for the first time IRT modeling was used for scaling items into a total score as well as multiple sub-scores. Up until TIMSS-95, student performance was mostly reported in the percent correct metric. Like SIMS and SISS, TIMSS-95 used matrix sampling with multiple forms, but SIMS and SISS did not employ IRT modeling. TIMSS-95 ended up with 102, 151, 38, and 65 items by which to estimate country mathematics performance in Populations 1, 2, 3 (literacy), and 3 (calculus), respectively. In science the comparable numbers were 97, 135, 26, and 65 (physics).

PISA followed a similar approach both in having professional item writers develop the needed item pool but also in the use of rotated forms. In effect, both TIMSS-95/TRENDS and PISA have followed similar procedures in this area.

Scaling

Taking advantage of new psychometric developments at the time, TIMSS-95 produced score reports of student achievement through scales based on Item Response Theory (IRT). In particular:

> The scaling model used in TIMSS was the multidimensional random coefficients logit model as described by Adams, Wilson, and Wang

(1997), with the addition of a multivariate linear model imposed on the population distribution. The scaling was done with the *ConQuest* software (Wu, Adams, and Wilson, 1997) that was developed in part to meet the needs of the TIMSS study (Adams, Wu, & Macaskill, 1997, p. 111).

Due to the limited number of items taken by each individual student given the constrained amount of time permitted for testing any single student, the IRT scaling was combined with conditioned multiple imputation methodology in order to assure reliable estimates. These procedures used in the psychometric context are also called plausible values (Mislevy, 1991). Typically, five estimates are generated for each student's proficiency. From these five estimates, an average provides an estimate of student achievement, and from the variance of those five plausible values an estimate of the standard error of measurement can be generated. The full statistical development of these procedures (processes) is described in the TIMSS-95 Technical Volume II, which also offers the following caution:

> Further information on plausible value methods may be found in Mislevy (1991), and in Mislevy, Johnson, and Muraki (1992). The proficiency scale scores or plausible values assigned to each student are actually random draws from the estimated ability distribution of students with similar item response patterns and background characteristics. The plausible values are intermediate values that may be used in statistical analyses to provide good estimates of parameters of student populations. Although intended for use in place of student scores in analyses, plausible values are designed primarily to estimate population parameters, and are not optimal estimates of individual student proficiency (Adams, Wu, & Macaskill, 1997, p. 111).

PISA uses the same methodology for scaling its assessments.

Ranking Countries

As we have said before in the introduction to this chapter and elsewhere, the original purpose behind launching quantitative international assessments was to obtain quantitative data so as to objectively compare countries. But what did the researchers mean by the word compare? The obvious comparison is in terms of average student performance.

> Because the IEA research ventures were launched during the post-Sputnik period, our cross-nationally comparative study was inevitably affected by the climate created by the race for superiority in science and technology. As early as the 1950s many Americans believed that the fight for world supremacy had to be fought in classrooms by increasing the number of students who took science and by raising educational standards. The

> National Defense Education Act was passed in the fall of 1958 for the purpose of strengthening the infrastructure of American technology. Massive resources were made available in the United States for programs to upgrade mathematics and science curricula and instruction. When the IEA study was launched, what in the minds of some academics was perceived as a major exercise in basic research was perceived by others as an international contest in mathematics. Now, it would at last be possible to find out which country scored highest (Husén, 1979, p. 379).

Given the description in the first part of this chapter of what has been accomplished over the past sixty years, we might conclude that the founding fathers accomplished what they wanted; or did they? The answer might not be the obvious one, as Husén noted:

> The tests were not devised primarily in order to make total score comparisons between countries possible and certainly not as yard sticks for an "international contest". The mere fact that algebra and geometry items were included in the tests for the 13-year level in spite of the fact that these topics were not dealt with in some countries should discourage national comparisons. The construction of the tests was guided by the hypotheses advanced and the tests are to be used primarily for comparisons between school systems both within and between countries in relation to these hypotheses. Thereby, subscores play as important a role as do variables like student's opportunity to learn the items and emphasis put upon the topics in the instruction (Husén, 1979, p. 26).

As implied in Husén's comment, it is also possible to compare countries in terms of the typical curricular experiences of students (OTL) in both mathematics and science. A key distinguishing feature across the different studies is the extent to which they address each of these two ways to answer the question: compare countries on the basis of what? Some sixty years after that original foundational meeting in Hamburg, Germany, the dominant driving force has resulted in the answer – on the basis of the average level of student performance.

But it was not always that way. Both the longitudinal version of SIMS – focusing on classroom processes – and the original TIMSS-95 – focusing on country textbook and standards documents as well as policy documents to characterize the full K–12 curriculum and the associated curricular structure and relevant policy formulation – pushed the envelope to focus on curricular comparisons as well as performance. Those studies saw curricular comparisons certainly as equally, if not more, important in answering the original question about the basis for making comparisons among countries.

The tension between the two approaches to comparing nations was already present during the ramp up to the First International Mathematics Study (FIMS). As Torsten Husén, the President of IEA at the time noted, "The fact that these comparisons are cross-national should not be taken as an indication that the primary interest was, for instance, national means and dispersions in school achievements at certain ages or school levels" (Husén, 1967a, p. 30). He further states, "Of great interest to us have been questions to do with the subject matter and the training of teachers of mathematics" (p. 32). So it is very clear that from the beginning international studies of educational achievement were equally focused on the inputs of schooling as well as the outputs.

To accomplish the country comparisons for the horse race, which this chapter focuses on, it is important to note that the field has moved beyond the total test score defined as the number correct or the percent correct, which was the case through SIMS (SIMS, however, reported only multiple test scores), to some very sophisticated psychometrics that began around the same time. The Danish mathematician Georg Rasch, who was an attendee at the original meeting in Hamburg, had advanced a new approach to scaling test performance based on a mathematical model called the logistic function.[1] This model solved a big problem by providing a more objective scaling method that was both person and item free. This was introduced for the first time in the Third International Mathematics and Science Study.

That only provided more sophisticated psychometrics that allowed the studies to create objective measurements but the question still remained; comparisons on what basis? Put differently, do we want an overall indication of performance on the mathematics domain as a whole or do we want to make comparisons among countries on subparts of the overall domain?

Mathematics, although a single discipline, is composed of various subfields, including algebra, geometry, number theory, calculus, and analysis. The more interesting comparisons could well be how countries compare on these various subdomains and, if so, do those differences reflect differences in the curriculum? This is even truer in assessing science, as this area itself is composed of at least four major disciplines, biology, chemistry, physics, and earth science, each with multiple subdivisions.

The above subdomains can be even further divided, yielding greater specificity in terms of student performance, especially as they might be related to curricular differences across countries. There are some who

---

[1] The Rasch model is equivalent to a 1-parameter IRT model.

believe that the total test score comparisons should not even be provided as they encourage uninformed judgments about the meaning of such country differences. This occurs at each cycle of these studies when a new "winner" is identified as the highest performing country or in some cases a city such as Shanghai in the recent 2012 PISA mathematics study.

The winners over recent cycles have included: Singapore, Japan, Finland, and Shanghai. This adulation and seeking of the Holy Grail by which all who drink from the chalice will somehow become winners too is an unsophisticated and naïve response and would have the founders turning over in their graves. It is best said in Husén's own words following the FIMS study:

> At the first, great efforts were made to play down the "horse race" aspect by referring to the fact that countries had different curricula. We pointed out that differences in average performance between countries could not without great reservations be interpreted as reflecting differences in the efficacy of mathematics education because of the impact of social and economic differences on student competence. Furthermore, the structure and selectivity of the systems played an important role. Although 13-year-olds in England and Germany, who had transferred to academic, selective secondary schools, had already been confronted with algebra and geometry, this was generally not the case in Sweden and the United States, the two countries with the lowest average performances at that age level. Despite efforts to point out such causes for differences in national scores, the outcry was tremendous in both these countries (1979, p. 380).

While this was a caution and worry in other IEA studies, they nonetheless produced a total test score often with subdomain scores as well. This issue rose to an elevated level of attention in SIMS. The study refused to provide the league tables based on the total score so as to avoid this kind of response. They only provided country comparisons with respect to five subdomains: algebra, arithmetic, geometry, measurement, and descriptive statistics.

The SIMS cross-country comparisons found in Table 4.2 presented z-scores so as to facilitate the comparison of performance across subtests. As an example of the type of comparisons drawn, we quote the report itself; "For example, students in five systems performed best on the Arithmetic subtest, while those from six other systems had their poorest performance on that same set of items. Similar differences occurred on all five subtests" (Robitaille, 1989, p. 123).

However, as we turn to TIMSS-95, things changed again, with respect to the question – on what basis do we compare the countries? The leaders

Table 4.2 *Subtest comparisons: SIMS population A (z-scores)*

	Arithmetic	Algebra	Geometry	Measurement	Descriptive statistics
Belgium (Flemish)	1.0	1.1*	0.2•	0.8	0.4
Belgium (French)	0.9*	0.7	0.2	0.7	−0.3•
Canada (British Columbia)	1.0*	0.6	0.1•	0.1•	0.7
Canada (Ontario)	0.6*	−0.1•	0.2	0.0	0.2
England and Wales	−0.3•	−0.3*	0.5	−0.2	0.6*
Finland	−0.6•	0.1	0.2	0.0	0.4*
France	1.0	1.4*	−0.4•	1	0.3
Hong Kong	0.6*	0.0	0.1•	0.2	0.0
Hungary	0.9	0.8	1.4*	1.2	−0.6•
Israel	0.0	0.1*	−0.6•	−0.5	−0.3
Japan	1.2•	1.9	2.1*	2	1.8
Luxembourg	−0.7	−1.4	−2.0•	−0.1*	−1.9
Netherlands	1.1	0.9•	1.3*	1.2	1.2
New Zealand	−0.6•	−0.4	1.3*	−0.6•	0.3
Nigeria	−1.2*	−1.2*	−1.9	−2.2•	−1.9
Scotland	−0.1	0.0	0.6*	−0.3•	0.5
Swaziland	−2.4•	−2.0	−1.3*	−1.7	−2.0
Sweden	−1.2•	−1.2•	−0.3	−0.2	0.2*
Thailand	−1.0	−0.6•	−0.3*	−0.3*	−1.0•
United States	0.1	−0.1	−0.4	−1.1•	0.4*
International Mean	50	43	41	51	55
SD	7.6	8.8	8.1	9.1	9.3

Note. * Best subtest performance for that system; • Poorest subtest performance for that system. Adapted from: "Students' Achievements: Population A" by D. F. Robitaille, 1989, In D. F. Robitaille and R. A. Garden (Eds.), *The IEA Study of Mathematics II: Contexts and outcomes of school mathematics*, p. 124. Copyright 1989 by International Association for the Evaluation of Educational Achievement (IEA). Publisher: Pergamon Press.

of the study, several of whom were part of the SIMS leadership, considered following in the footsteps of SIMS by not reporting total-score league tables. As the study expanded from a third mathematics study at eighth grade to a study of fourth, eighth, and twelfth grade, as well as including science under the encouragement of the US National Center for Educational Statistics (NCES), those ideas had to be reconsidered. The involvement of the US government and the corresponding funding for the International Center came with a clear mandate that included the creation of country comparisons on a total test score created by IRT modeling. This was done so as to parallel the US National Assessment of Educational Progress (NAEP), which also occurred at fourth and eighth grade and provided total test score comparisons of states.

TIMSS-95 did produce the standard "horse race" tables as does PISA. In response, Leigh Burstein (one of the original leaders of TIMSS-95), in commenting on the study as it was being developed, pointed to this as a turning point in IEA studies: from the research-oriented earlier studies to the more bureaucratic government-type study focusing on country comparisons (Burstein, 1991, 1993). He was right, but only partially. TIMSS-95 did, as a result of the involvement of the US government, become more government-like in terms of the sampling requirements, test construction, scaling, and reporting of the results. But he was wrong in assuming the study was not going to be driven by researchers with a clear set of research questions; unfortunately, he never saw that his concerns did not occur, as he died before the study was completely designed. The study, as it was more fully developed, did have a research focus and was driven by a conceptual framework that linked all parts of the study, including the curriculum study, the test through its blueprint, and the teacher and student questionnaires. The research-related activities were funded by the US National Science Foundation.

TIMSS-95 did provide subtest scores. The study produced tables comparing countries on six subdomains in mathematics and five in science. Although somewhat similar to SIMS, the subdomains were more refined: fractions and number sense; geometry; algebra; data representation analysis and probability; measurement; and proportionality. The choice of these subdomains was based on the TIMSS-95 curriculum analysis described in Chapter 7. At eighth grade the science categories were the traditional four as used in SISS but also included one other subtest that included environmental issues and the nature of science.

Follow-up analyses done by the US team created and compared countries on still further refined subdomains (see Chapter 5). Given the extensive nature of the TIMSS-95 curriculum study and the direct linkage of that analysis to the development of the tests through the frameworks, this was possible and desirable as the two types of data could be linked at a level of specificity that more closely approximates the level at which learning takes place (Schmidt, McKnight, Cogan, Jakwerth, & Houang, 1999; Schmidt, McKnight, Houang, Wang, Wiley, Cogan, & Wolfe, 2001).

For example, one would expect the amount of time related to the study of adding fractions without a common denominator to be related to a test of adding and subtracting fractions more than you would the amount of time spent on the study of the topic area called whole number and fractions. This approach led to country comparisons at eighth grade on twenty subtests in mathematics (including such subdomains as: decimals and percents; perimeter, area, and volume; 3D-geometry and transformations), and seventeen in science (including the subdomains: chemical changes; structure of matter; life cycles and genetics). At fourth grade there were fourteen subtests in mathematics and fifteen in science. The discussion of the results of these subtest score analyses produced a fascinating set of conclusions about the nature of cross-country comparisons and what they mean and do not mean. This is the topic of the next chapter.

CHAPTER 5

Relating Assessment to OTL
Domain-Sensitive Testing

If the proverbial man-on-the-street knows anything today about international assessments of education it is that these assessments are succinctly summarized in the media with a ranked list based on the average scores of participating countries with the top scorer listed first. Sometimes, only the top scorer is mentioned along with the ranks of one or two other countries of interest. Despite the relatively successful efforts of the early pioneers to avoid the spectacle of a "cognitive Olympics," this is the sound bite metaphor that has developed and, after TIMSS-95, has prevailed.

Around the world, most informed citizens likely do not recognize the acronyms of the most recent international comparisons, e.g. TIMSS, PISA, PIRLS, ICCS, but would likely know where their education system ranked compared to others on the most recent assessment of mathematics, science, or reading. Such a vague familiarity may be compared to a general awareness that one's Olympic team won ten gold medals in the competitions without knowing which specific events they won.

Unfortunately for the general public, news reports of these studies with their focus on the total score do not tend to provide any further insight into the state of education in any particular country, including their own, other than to point out rank orderings. Rank orderings on a total score with the implicit presumption that this also reflects the quality and efficiency of the education systems is exactly what the founding fathers of modern international comparative education studies were trying to avoid. Why were they so concerned about this? Husén (1974) stated it this way:

> At the beginning the main emphasis was on description, i.e. on the construction of measuring instruments that could be employed in international comparisons. This led some outsiders to regard the survey as an international 'cognitive olympics' or a kind of horse race (cf. Findley 1971). It was evident, however, that achievement 'profiles' could be used to elucidate strong and weak points in a particular system (p. 17).

Primarily, they were interested in learning something about the nature of education in each country and a total score – while efficient for ranking countries – does not really provide insight into the nature of schooling either from a structural or instructional perspective.

Academics and researchers have been able to make use of total scores from the IEA studies and PISA together with other information gathered in those studies about systems, school characteristics, teachers, and student background (see Chapter 8 related to students' socioeconomic background) to identify important relationships. For example, in a re-analysis of results across three early IEA studies that assessed reading, mathematics, and science, Coleman (1975) concluded that schools made a contribution to students' performance, at least in mathematics and science. In his analysis, achievement in reading appeared to be largely a function of the quality of the home environment.

As important and intriguing as such findings are they do little to move beyond the simple and rather commonsense notion most have developed from their own educational experiences: Schooling really does matter and what students study in their classrooms makes a difference to what they know and are able to do. As long as the goal of international comparative assessments is to provide the man-on-the-street with a sound bite about the relative standings of one's educational system in a general area of school such as mathematics or science, a total scaled score is admirably efficient.

However, if such studies – which require a substantial investment of resources – are to do anything more, i.e. to be more informative and valuable to education policy makers and educators, then something more than a total scaled score is needed. In order to be useful to the various stakeholders for whom education is a matter of professional effort, international comparative assessments need to provide scores that move beyond a total mathematics or science score. They even need to move beyond the highly aggregated, broadly labeled sub-areas such as algebra or physics to provide scores that reflect to a greater degree the specific areas of what students have been studying in their classrooms (Schmidt, Jakwerth, & McKnight, 1998). Students do not learn mathematics in some general sense and they do not even learn algebra; what they do learn, for example, is the particular concept of a linear equation and how to solve it. Put metaphorically, as the grain size of the sampled content domain becomes smaller the more useful the results are to countries and their defined curriculum standards and policies.

Over the past twenty years, the major mathematics and science assessments, TIMSS-95, TRENDS, and PISA, have not only produced a total scaled score but also a small number of sub-scores defined by several broad content areas. The TRENDS assessments for nine-year-olds (fourth grade) report scores in three content subdomains: number; geometric shapes and measures; and data display. Four content subdomains are reported for thirteen-year-olds (eighth grade): number; algebra; geometry; and data and chance (Mullis, Martin, & Foy, 2008; Mullis, Martin, Foy, & Hooper, 2016). Similarly, the PISA mathematics assessment of fifteen-year-olds provides scaled scores for four content domains: change and relationships; quantity; shapes and space; and uncertainty and data (OECD, 2004; OECD, 2013a). Although this would appear to be a step in the direction of providing more information, the reality is that the scaled scores for these broad content subdomains differ little from the overall total scaled score.

To illustrate, in the 2015 TRENDS the top scoring country at eighth grade, Singapore, had a total score of 621 with scores in the sub-areas ranging from a low of 617 in both the areas of geometry and data and chance to a high of 629 for number. This pattern was evident for most of the other participating countries, with a few having a somewhat larger variance for the subdomain scores. The United States had a total score of 518 with a low of 500 (geometry) and a high of 525 (algebra). This has not surprised some researchers, who have pointed out that this is what is to be expected from the shallow item sampling from broadly defined content domains, which is true not only for the total domain of mathematics but for the broad subdomains typically used in these studies as well, and the selection of items to maximize the reliability of a normative scaled score (Schmidt, Jakwerth, & McKnight, 1998).

An Alternate Approach to Scaling Subdomains: Percent Correct Scores

In addition to the IRT scaled scores in the sub-areas, TRENDS-2015 also reported the percent correct. Although the percent correct scores correlate virtually perfectly with the scaled scores at the country level, the percent correct metric has a more accessible and applicable meaning, especially to policymakers and the public. Singapore's overall scaled score was 621; other than being the highest overall score of any country, the meaning of this number is not apparent as it is only comparable to the average score of 500, which is the international mean.

However, Singapore's overall average percent correct of 74 percent on the mathematics test implies that a typical Singapore eighth-grade student responded correctly to 74 percent of the TRENDS items. Comparing the percent correct across the sub-areas provides further insight. However, this needs to be done cautiously because such comparisons depend on the difficulties of the particular items used to sample the domains. Interpreting them depends on the viability of the assumption that they are random samples of the domains of items.

Singapore students' best performance was in the area of number, in which they, on average, responded correctly to 80 percent of the items. Their lowest score, 68 percent correct, was in the area of geometry. The percent correct metric suggests a greater connection to schooling, with its interpretation of the proportion of items in a given subdomain of mathematics that students are able to successfully complete.

Australia provides a more intriguing example. Australia's overall average percent correct was 45 percent, a bit higher than the international average of 41 percent. However, their performance ranged from a high of 55 percent in the area of data and chance to a low of 35 percent in the algebra area, which was below the international average of 37 percent. The low international average suggests that this area had items that were more challenging for students in many countries. The low percent correct for Australia's students suggests that students were not familiar with many of those items; perhaps they have only recently begun to study the type of mathematics needed to respond correctly to them. The use of percent correct scores, however, has it limits. Scaled scores can be equated from one assessment to the next, but percent correct scores will fluctuate with the particular item samples used. That is, trends are difficult with percent correct. The use of the two scaling procedures together might provide a broader context in which to interpret the results.

Domain-Sensitive Assessments

This leads us back to the question posed in Chapter 4: Do we want an overall indication of performance on the whole of the mathematics or science domain as defined for a specific population or do we want to make comparisons among countries on smaller subsets of the overall domain?

Students encounter a mathematics or science curriculum in which there are numerous topics to be learned. They may master some content, but not other content, leading to a profile of performance across the 15–30 (mathematics) or the 30–60 (science) content areas a typical sixth to ninth grade student in most countries encounters (for mathematics see Figure 7.2) (Schmidt, McKnight, & Raizen, 1997). When individual areas of competence are aggregated to a total score, the interpretation of the percent correct loses any precise meaning. Such a score mostly reflects a general propensity or aptitude for mathematics or science and, as such, is not a helpful indicator of what the student has learned in school, which is a central purpose of these international assessments. In fact, such a total score is more likely to be a reflection of the social class (SES) and the motivation level of the student together with his/her propensity to do well in the area of mathematics or science, thus having little to do with schooling.

If ranking countries is the goal of the assessment then the total score is appropriate. However, if the purpose is to determine whether students learn what they have been taught so as to monitor the impact of schooling in a country, or to compare schooling across countries looking at differences in curriculum, then such an approach is not the most helpful. Neither participating countries nor comparative education researchers looking for meaningful comparisons of schooling are likely to have much light shed on their concerns by a simple total score.

For IEA studies, developing assessments in specific areas of mathematics or science, defined at a level reflective of instructional units typically taught at the designated grade level across countries, seems the most reasonable way to proceed. We term this curriculum-sensitive assessment. Consequently, the country performance would be expected to reflect the country's curricular emphasis profile with respect to the content areas in the curriculum: not only its coverage but also how much emphasis it has received.

If the curriculum of a country, for example, specifies that a sizable proportion of the eighth-grade curriculum should cover congruence and similarity, while in another country the same topic received very limited coverage, then the comparison of the performances of the two countries on an assessment focused on that subdomain would provide meaningful information about schooling and in particular the curriculum standards, thus providing curriculum-sensitive measurement. The converse is also important: If two countries devote the same curricular emphasis to a topic but achievement is high in one and low in the other, one needs to search for explanations, such as quality of instruction. This is why Husén raised the issue of comparing countries not only on performance but also in terms of the curriculum. The more specific the subdomain of mathematics on which the country comparisons are based, the more researchers and policymakers can learn about schooling; a key presumed goal of international assessments. The same logic can apply to regional, school, or classroom comparisons within a country. The next section of this chapter provides a look at the TIMSS-95 results where an approach to such comparisons was made.

In contrast, the domain for PISA reflects the definition of mathematical or scientific literacy. The mathematics framework developed for PISA in 2012 was specified as a model of problem solving that included several competencies. Literacy is not defined relative to the school curriculum but rather to the desired competencies defining literacy. Consequently, we use the term competency-sensitive assessment, which parallels curriculum-sensitive assessment, since "literacy" is not taught as a separate topic in the mathematics or science curriculum but reflects the cumulative exposure of students across their schooling up to the age of 15. For example, PISA could develop assessments defined at sublevels of the four content areas such as using computer simulations in problem solving. Similarly, assessment sub-scores could be developed in solving problems related to personal health issues.

For PISA and IEA studies, both types of domain-sensitive assessments would result in scores more closely related to schooling, i.e. the learning experiences students have been exposed to in their schools.

Toward Curriculum Sensitive Measurement in IEA Studies

In TIMSS-95, researchers were able to construct percent correct scores at the country level for a large number of content areas that reflected aspects of the curriculum at a given grade level. These were reported and discussed most fully in *Facing the Consequences* (Schmidt, McKnight, Cogan, Jakwerth, & Houang, 1999). Altogether, fourteen mathematics curricular content areas and fifteen science curricular areas were developed for fourth grade (nine-year-olds). At eighth grade, twenty mathematics areas and seventeen science areas were developed from all the TIMSS-95 items.

The percent correct score metric defined for highly specific topic areas can be more readily interpreted with relation to schooling and more directly to the curriculum. Besides the issue of clarity in communicability, such highly specific subdomain scores are only doable in a practical sense using the percent correct matrix. The matrix sampling design of the assessment and the sampling of students for all the international assessments precludes the construction of percent correct scores for individual students. Providing a percent correct score for all students would require that each student respond to many more items, with the consequence that the assessment time would need to be substantially increased. Nonetheless, given current sampling procedures and the use of the item booklet rotation design (such as the duplex design), percent correct scores at the country or school levels are possible if items are selected to represent a number of more narrowly defined curricular areas.

Test-Curriculum Match Analysis in IEA Studies

The appropriateness (validity) of the student assessment for students in each participating country has been a concern from the beginning of IEA studies. Subject matter experts were provided with copies of all the proposed student assessment items and asked to indicate whether each item was appropriate for the target-grade students in their education system. The precise phrasing of this question differed from study to study but the information gathered from this exercise was used to judge the appropriateness (validity) of including items on the assessment across all systems. Results from this analysis were not reported in detail for some studies but were mentioned at some point in the reports to support the validity of making comparisons of student achievement across systems.

The process was updated and expanded with TIMSS-95. As in prior studies, country experts were asked if each item from the relevant student test "was in the country's intended curriculum" for students at the appropriate grade level, e.g. fourth or eighth grade (Beaton et al., 1996b, p. B-3). Percent correct scores were then generated for all countries based on the subset of items each country's expert had selected as part of the curriculum for their students. The result of this was that "half of the countries indicated that items representing 90% or more of the score points ... were appropriate, with the percent ranging from 100% in Hungary and the United States to 47% in Greece" (Beaton et al., 1996b, p. B-2).

Comparing the average country percent correct scores on the different sets of items defined by countries' experts as appropriate for the students in their country, the conclusion was "that different item selections do not make a major difference in how well countries perform relative to each other ... The relative performance of countries on the various items selected did vary somewhat, but generally not in a statistically significant manner" (Beaton et al., 1996b, p. B-5). The wording regarding these results was nearly identical in Appendix B for both the mathematics and science Test Curriculum Matching Analysis (TCMA) reports for primary and middle school (Mullis et al., 1997; Martin et al., 1997; Beaton et al., 1996a; 1996b).

An additional note in the middle school science report appendix stated, "Although there are some changes in the ordering of countries based on the items selected for the TCMA, most of these differences are within the boundaries of sampling error"(Beaton et al., 1996a, p. B-5). However, a footnote on the first page of the appendix notes that "[b]ecause there also may be curriculum areas covered in some countries that are not covered by the TIMSS [TIMSS-95] tests, the TCMA does not provide complete information about how well the TIMSS [TIMSS-95] tests cover the curricula of the countries" (Beaton et al., 1996b, B-1). A nearly identical footnote appears in Appendix F of the most recent (as of the writing of this chapter) TRENDS (Mullis et al., 2016).

An examination of Table 5.1, an adaptation of Table B.1 from the 1995 TIMSS middle school mathematics report, clearly supports the conclusions stated in the reports on the TCMA.

Curriculum-Sensitive Test Score Analysis in TIMSS-95

Using early results from the TIMSS-95 curriculum analysis and the field trial data, Jakwerth (1996) constructed a comprehensive test-curriculum matching analysis. Rather than using experts rating, as in the previous section, of whether an item was appropriate for students at a particular grade, this analysis relied upon the exhaustive coding of each country's K–12 curriculum standards (guides) (see Chapter 7) to determine the match with each item that appeared on the field test student assessments. Average percent correct across all field trial items was calculated for the seventeen countries providing sufficient field trial data. In addition, average percent correct scores were computed as defined by the curriculum standards from each of the seventeen countries. Thus each field trial country had eighteen average percent correct scores: one over all field trial items and one for each of the seventeen country-defined tests.

Similar to the TCMA results included in the official TIMSS-95 report volumes, Jakwerth (1996) noted that "[w]hen scores and ranks are averaged over all the specially-constructed tests, little difference is seen from the original field-trial scores and ranks. Average differences in passing rates on the field-trial instrument and passing rates on other tests are only about 3%, and all country ranks are nearly identical" (p. 175).

However, this assessment and curriculum alignment analysis was not the only test definition explored in the analysis. Making use of the TIMSS-95 Mathematics Curriculum Framework created for the in-depth curriculum analysis (Schmidt, McKnight, Valverde, Houang, & Wiley,1997), Jakwerth (1996) defined twenty-nine framework-defined topic-specific subtests and compared the average percent correct for each

Table 5.1 *Average percent correct on country-defined sets of appropriate eighth-grade items*

Test-Curriculum Matching Analysis Results - Mathematics - Upper Grade (Eighth Grade*)
Average Percent Correct based on Subsets of Items Specially Identified by Each Country as Addressing Its Curriculum (See Table 6.3 for corresponding standard errors)

Country	(Number of Score Points Included)	Singapore	Korea	Hong Kong	Belgium (Fl)	Czech Republic	Slovak Republic	Switzerland	Austria	Hungary	France	Slovenia	Russian Federation	Netherlands	Bulgaria	Canada	Ireland	Belgium (Fr)	Australia	Israel	Sweden	Germany	New Zealand	Norway	England	United States	Denmark	Scotland	Latvia (LSS)	Spain	Iceland	Greece	Romania	Lithuania	Cyprus	Portugal	Iran, Islamic Rep.	Kuwait	Colombia	South Africa
Singapore	79	79	80	79	79	79	79	80	80	79	79	79	81	79	80	79	79	79	79	79	79	79	79	79	81	79	80	80	79	79	80	79	79	79	80	79	79	79	80	79
Japan	73	73	74	73	73	73	73	75	74	73	73	73	74	73	73	73	73	73	73	73	73	73	73	73	75	73	75	74	73	74	74	73	73	73	73	73	73	73	74	74
Korea	72	71	73	72	72	72	72	73	72	72	72	72	73	72	72	72	72	72	71	72	71	72	72	72	72	72	73	72	72	71	72	72	72	72	71	72	72	71	72	72
Hong Kong	70	70	71	70	70	70	70	71	71	70	70	70	71	70	70	70	70	70	70	70	70	70	70	70	70	70	70	70	70	70	70	70	70	70	70	70	70	70	70	70
Belgium (Fl)	66	65	67	66	66	66	66	68	67	66	66	66	67	66	66	66	66	66	66	66	66	66	66	66	68	66	67	67	66	67	67	65	66	66	66	65	66	66	66	65
Czech Republic	66	65	67	66	65	66	66	68	66	66	66	66	67	64	66	66	66	66	66	66	66	66	66	66	67	67	67	66	65	67	68	65	66	66	65	64	66	66	66	65
Slovak Republic	63	63	64	63	63	63	63	65	63	63	63	63	63	63	63	63	63	63	63	63	63	63	63	63	64	63	63	63	62	63	64	63	63	63	63	63	63	63	63	63
Switzerland	62	61	63	62	61	62	62	63	63	62	61	62	63	62	62	62	61	62	61	62	61	62	62	62	63	62	63	62	62	62	63	62	62	62	62	62	62	61	62	62
Austria	62	61	63	62	61	62	62	63	62	62	61	62	63	61	62	62	61	62	61	62	61	62	62	62	63	62	63	62	62	62	63	62	62	62	62	62	62	61	62	62
Hungary	61	61	62	61	60	61	61	63	61	61	61	61	62	60	61	61	61	61	61	61	61	61	61	61	63	61	62	61	61	61	62	61	61	61	61	61	61	61	61	61
France	61	61	62	60	60	61	61	63	62	61	61	61	62	61	61	61	61	60	61	60	60	61	61	61	63	61	62	61	61	61	62	60	61	61	60	60	61	60	61	61
Slovenia	61	61	62	60	60	61	61	62	60	60	60	60	62	61	61	61	60	60	60	60	60	60	60	60	61	61	62	60	60	61	61	60	60	60	60	60	60	60	62	61
Russian Federation	60	60	61	60	58	59	59	61	60	60	60	59	61	58	60	60	59	59	59	59	60	60	59	59	61	60	60	60	60	61	61	59	59	60	58	59	60	60	59	58
Netherlands	60	59	60	58	58	59	59	61	60	59	59	59	61	59	60	59	58	59	58	60	59	60	58	58	61	60	60	60	59	61	61	58	59	58	58	58	59	59	59	58
Bulgaria	59	59	59	58	57	58	58	60	59	58	58	58	59	58	60	59	58	58	58	58	59	58	58	58	62	59	60	60	58	60	60	58	58	58	58	58	59	58	58	58
Canada	59	58	59	58	57	58	58	59	58	58	58	58	59	57	58	59	58	57	57	58	58	58	58	57	59	58	58	58	58	59	59	58	58	58	57	57	58	58	58	58
Ireland	58	58	58	58	57	58	58	59	58	58	58	58	59	57	58	58	58	58	57	58	58	58	57	57	59	58	58	58	58	58	58	57	58	58	57	57	58	57	57	57
Belgium (Fr)	58	57	60	57	57	58	58	59	58	58	58	58	59	57	58	58	57	58	56	58	58	58	57	57	61	58	58	58	57	59	59	57	58	58	57	56	58	57	58	57
Australia	57	57	58	57	56	57	57	58	57	57	57	57	58	57	57	57	57	56	57	57	57	57	56	56	59	57	58	57	57	58	58	57	57	57	57	56	57	57	57	57
Israel	55	55	56	55	54	55	55	57	55	55	55	55	56	55	55	56	55	54	54	55	55	55	54	54	57	55	56	55	55	55	56	55	55	55	55	54	55	55	54	54
Sweden	54	54	55	55	53	54	54	55	54	54	54	54	55	54	54	55	54	52	54	54	54	54	54	54	55	54	55	55	54	55	55	54	54	54	54	54	54	54	54	53
Germany	54	53	54	53	53	53	53	55	54	53	53	53	54	53	53	54	53	52	53	53	53	53	53	52	55	53	55	54	53	54	55	53	53	53	53	53	53	53	53	53
New Zealand	54	54	54	53	52	53	53	54	54	53	53	53	54	53	53	53	53	52	52	53	53	56	53	53	56	54	54	54	53	54	55	53	53	53	52	52	54	53	53	52
Norway	54	53	54	53	52	53	53	54	53	53	53	53	54	53	54	54	53	52	52	53	53	53	54	53	54	54	55	53	53	55	54	53	53	53	53	53	54	53	53	53
England	53	52	53	52	51	52	52	54	52	52	52	52	53	51	52	53	52	50	51	52	51	53	51	52	57	52	53	52	52	53	54	51	52	52	51	51	53	52	52	52
United States	52	52	52	51	51	52	52	53	52	52	52	52	53	50	52	52	51	49	51	52	51	52	51	51	55	53	52	52	51	53	53	51	51	52	51	50	52	52	52	51
Denmark	52	51	52	51	51	51	51	53	52	51	51	51	53	50	52	52	51	49	51	52	51	52	51	51	55	53	52	52	51	52	53	51	51	52	50	49	52	52	50	51
Scotland	51	51	52	51	51	51	51	53	51	51	51	51	52	50	51	52	51	50	51	51	51	51	51	51	53	53	51	52	51	52	52	50	51	51	50	49	51	52	51	51
Latvia (LSS)	51	50	52	51	49	51	51	52	52	50	50	50	53	49	51	51	50	50	50	51	50	51	50	49	53	51	51	52	50	52	53	49	50	50	49	49	51	50	50	50
Spain	50	49	50	49	48	49	49	50	49	49	49	49	50	48	50	50	49	49	49	49	49	50	49	48	53	50	51	51	49	50	50	48	49	49	48	48	50	50	49	50
Iceland	49	49	49	49	48	49	49	50	49	49	49	49	50	48	49	49	49	48	48	49	48	50	49	47	50	50	49	49	49	50	50	47	49	49	48	48	49	49	49	48
Greece	49	49	50	49	48	49	49	51	49	48	48	48	50	47	50	49	48	49	47	49	48	50	48	47	50	50	50	48	48	50	51	48	48	49	48	47	49	48	49	48
Romania	49	49	49	48	47	49	49	49	48	48	48	48	50	47	48	49	48	49	47	48	48	49	48	47	50	50	49	49	48	50	50	48	48	48	48	47	49	49	49	48
Lithuania	48	48	48	48	47	48	48	49	48	48	48	48	49	47	48	48	48	47	47	48	48	49	48	47	50	50	49	48	48	50	48	47	48	48	48	47	48	48	49	48
Cyprus	48	47	49	47	47	47	47	47	47	46	46	46	48	48	47	47	47	47	46	47	47	47	47	45	45	48	47	47	47	48	48	47	46	47	47	46	47	47	47	47
Portugal	43	42	43	43	42	42	42	45	43	42	42	42	44	41	43	43	42	41	42	43	42	43	42	43	45	43	44	44	43	44	44	42	43	43	42	41	43	43	43	41
Iran, Islamic Rep.	38	38	39	38	37	38	38	39	38	38	38	38	39	38	38	38	37	38	37	38	37	38	38	38	39	39	39	38	38	39	39	38	38	38	37	38	38	38	39	37
Kuwait	29	29	30	29	29	29	29	31	29	29	29	29	30	28	30	29	29	29	29	30	29	30	30	30	31	30	30	30	29	30	30	29	29	29	28	28	30	30	30	28
Colombia	29	29	30	29	29	29	29	31	29	29	29	29	30	28	30	29	29	28	29	29	29	31	29	29	32	29	30	29	29	30	31	29	29	29	28	30	30	29	29	28
South Africa	24	24	24	24	24	24	24	25	24	24	24	24	24	23	24	24	24	24	23	24	24	24	24	24	25	24	24	24	24	24	24	23	24	24	24	24	24	24	24	23
International Average	55	55	56	55	54	55	55	57	55	55	55	55	56	54	56	56	55	54	54	55	55	55	55	54	58	56	56	56	55	56	56	55	55	55	54	54	56	55	55	54
Number of Score Points Included	162	144	148	150	140	150	152	133	147	162	140	151	126	116	119	147	145	138	154	159	127	155	145	150	130	162	135	125	161	158	133	76	143	155	124	152	147	140	133	129

Note. Source: "Mathematics Achievement in the Middle School Years: IEA's Third International Mathematics and Science Study," by A. E. Beaton, I. V. S. Mullis, M. O. Martin, E. J. Gonzalez, D. L. Kelly, & T. A. Smith, 1996, p. B-3. Copyright 1996, International Association for the Evaluation of Education Achievement (IEA). Publisher: TIMSS & PIRLS International Study Center, Lynch School of Education, Boston College.

country on each of these subtests. In stark contrast to the relative invariance among the total scores and ranks on the standards-defined tests, across these topic-specific subtests over half of the field trial countries had the highest percent correct on at least one of the twenty-nine topic-specific tests. In addition, 35 percent of the countries demonstrated the lowest score on at least one of the topic-specific tests. Perhaps even more startling was the observation that two of the seventeen countries that had the highest score on at least one topic-specific test also had the lowest score on a different topic-specific test.

Schmidt, Jakwerth, and McKnight (1998) conducted a similar comprehensive analysis of the TIMSS-95 assessment data, together with results from the curriculum analysis, and offered one explanation for the stark contrast between the relative invariance in total test scores defined by alignment with standards and the substantial variance in sub-scores defined more precisely by specific content areas. They noted "that curricular effects on performance (or performance variability) may be 'washed out' when item performance is aggregated over many multidimensional topics" (p. 513). They compared twenty framework-defined topic-specific test scores and ranks for the forty-one countries participating in the TIMSS-95 eighth-grade assessment.

Figure 5.1 is a boxplot of the country ranks across all of the twenty eighth-grade mathematics topic-specific tests. They noted "that 17 of 41 [41 percent of] countries ranked in the top five for at least one [m]athematics topic [subtest]. Thirty-one countries [75 percent] had ranks on [m]athematics topics [subtests] that fell in at least three different quartiles for differing topics" (p. 516). Also noted was that the average over all countries of the difference between the country's high and low ranking was 18.

As much as variability was observed in country rankings across a set of topic-specific defined tests, even greater variability in eighth-grade student performance was evident at the item level. Of the forty-one TIMSS-95 countries, twenty-three, or 56 percent, "ranked first on at least one mathematics item" (Schmidt, Jakwerth, & McKnight, 1998, p. 516).

Returning to Total Test Scores

As efficient as a total scaled score may appear, the scaling problematically assumes a unidimensionality for all the items on the test. Unfortunately, this necessarily "ignores the multidimensionality of subject matter ... which tend[s] to be specific rather than global" (Schmidt, Jakwerth, &

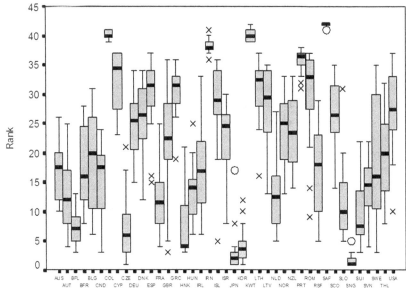

Figure 5.1 Distribution of country ranks across mathematics subtopics
Note. Republished with permission of Elsevier Science and Technology Journals, from "Curriculum Sensitive Assessment: Content *Does* Make a Difference," by W. H. Schmidt, P. M. Jakwerth, & C. C. McKnight, 1998, p. 515; permission conveyed through Copyright Clearance Center, Inc.

McKnight, 1998, pp. 509–510). Additionally, the aggregation of scores across a set of items introduces another threat to an assessment that is able to sensitively reflect differences in students' curricular experiences or OTL. Schmidt, Jakwerth, and McKnight point out that as "the level of aggregation increases, the specificity of commonalities portrayed decreases. At the highest, total score level, all that is left is general ability. Schooling and curriculum are often weakly relevant to general ability. ... High-level, global aggregation is thus fundamentally insensitive to curricular differences and unsuited for determining policies relevant to educational change" (Schmidt, Jakwerth, & McKnight, 1998, p. 519).

In-depth analyses of the student performance data, teacher-reported classroom instruction (OTL) data, and the official curriculum document data has made clear that the definition of "eighth-grade mathematics" is often not consistent across these three instantiations between countries, and, sometimes, such as in the United States, is not consistent within a country (Schmidt, McKnight, Houang, Wang, Wiley, Cogan, & Wolfe, 2001).

Ignoring the multidimensionality of a school subject such as mathematics, as found in the practice of using highly aggregated total scores – whether produced through models assuming unidimensionality (such as IRT and Rasch models) or ones reported in the percent correct metric – risks missing entirely the clear conceptual and effectual connection between what teachers have taught and what students have learned. The specificity of these measurements was noted well over thirty years ago by Airasian and Madaus (1983). They noted that schools "do not directly teach global constructs, but instead try to develop specific skills, introducing one skill at a time for the student to integrate with previously acquired skills. Hence, there are more apt to be differences between schools and programs at the specific objective level than at the total score level" (pp. 105–106).

Given these insights into the importance of constructing assessments that are sensitive to the differences in the curricula across many countries in doing international research, how should we now proceed? A more extensive consideration of this question is given in Chapter 11. Here, however, we can conclude by endorsing the closing recommendations made by Schmidt, Jakwerth, and McKnight (1998).

- The domains sampled by such tests cannot continue to sacrifice relevant specificity for broad, shallow sampling that produces a misleading appearance of test validity.
- Sufficient specificity needs to be included in test designs, even at the cost of some breadth, to ensure that policy-relevant aspects of curriculum and classrooms are captured so that test results are both sensitive to these factors and provide empirical information that pays more than "lip service" to helping guide educational change.
- High levels of aggregation are fundamentally distorting. They are prone to misleadingly simplistic interpretations that misguide educational policy. Extreme care should be taken in all aggregations to maintain specificity and educational validity in creating scores assessing commonalities of educational practice that may be the source of policy recommendations.
- Curricula vary and often "the devil is in the details.'. No achievement-comparison study should be undertaken that does not also collect relevant curricular and educational data and does not seek to create

accurate portraits of how educational opportunity is created, even when such portraits must be a mosaic of smaller, detailed pictures and stories rather than some more global, "magic bullet" solution.
- Arguments based on using larger, broader, more highly relevant achievement measures to increase reliability are spurious. There is no virtue in being reliably wrong, invalid, or irrelevant. Careful thought needs to be given to designs that can sustain reasonable reliabilities and usefully small standard errors without sacrificing curricular sensitivity.
- Curricular sensitivity is an issue in assessment design, implementation, interpretation, and use in achievement-comparison research, especially cross-national studies. Content truly does make a difference, especially in the search for relevant empirical findings to effectively guide educational policy (Schmidt, Jakwerth, & McKnight, 1998, pp. 524–525).

Performance and Opportunity to Learn

Chapters 6 and 7 on opportunity to learn (OTL) elaborate the role it has played in international comparative assessments. The inclusion of various OTL measures has served both to highlight the similarities and differences in systems' intended and implemented curricula as well as to provide a measure addressing issues of validity for the student assessment. This last rationale is our focus in this section; the use of opportunity measures to ensure the content validity of the assessments used to make comparisons across participating systems (Schmidt, Jakwerth, & McKnight, 1998).

The footnote included with each published report on the TRENDS Test-Curriculum Matching Analysis (TCMA) provides a hint of one of the validity challenges for any assessment: Does the assessment actually match the substance of the broader mathematics curriculum as intended, taught, and emphasized? An important point to be considered is that the curriculum for any particular grade does not exist in a vacuum; rather it is intimately connected to the curricula of prior grades and forms a foundation for what students will study in future grades. So, any particular item may be judged as appropriate for that grade but in one system that appropriateness stems from the fact that it represents the culmination of the prior grade's emphasis, while in another country the item is judged appropriate but has only been introduced at the very end of the grade in question.

Such differences are not inconsequential considerations in the interpretation of assessments that purport to be measures of performance in a specific academic area such as mathematics at a specific grade level.

Assessments that are intended to measure student progress (achievement) in some curricular area must be mapped to the intended and implemented curriculum students have experienced across the grades.

The great diversity in the intended mathematics and science curricula across all the participating countries in the TIMSS-95 curriculum analysis was the eye-opening take away from the two *Many Visions, Many Aims* volumes (Schmidt, McKnight, Valverde, Houang, & Wiley, 1997; Schmidt, Raizen, Britton, Bianchi, & Wolfe, 1997). Although great consensus was found for topics to include in the assessment at the two focal grades – grade 4 and grade 8 – this intersection across all systems masked a diversity of approaches to developing the curriculum, particularly across the K–8 years.

The picture that emerges from such a multi-year, multi-grade perspective on the curriculum tends to cast doubt on any strong validity claims made on the basis of a limited item-based curriculum survey focused only at the target grade, such as the TCMA conducted as part of the recent TRENDS studies.

A Final Thought

A total scaled score efficiently and effectively accomplishes what it is theoretically and technically intended to do: It implicitly ranks individuals and their aggregated groups, schools, and classrooms, and explicitly and publically ranks countries, according to their responses to a selected group of items. Consequently, the media and the general public can be excused for not seeing past the "horse race" ranking of those countries that participated in the assessment. Such scores do not, however, provide insight into the state of education in the various participating educational systems; i.e. the relative rigor or quality of the curriculum students have studied, the quality or focus of their classroom instruction, an indicator of how well students may have learned what they have been taught, or the particular strengths or deficiencies of the educational system. Despite any stated claims made, such insights are precluded by the very nature of the technical scaling requirements that produce the total scaled score. If we accept the notion that an assessment's validity is related to "the quality of the inferences made using evidence obtained for the purposes the test is intended to serve" (p. 110, Baker, Chung, & Cai, 2016), one may understandably be forgiven for expressing concerns over the validity of the total scaled scores generated in modern international comparative assessments of education.

CHAPTER 6

The Evolution of the Concept of Opportunity to Learn

The histories of international assessments and the measurement of opportunity to learn, including its conceptual development, are inextricably intertwined. As noted in Chapter 1, the founding fathers of the modern era of international comparisons knew that substantial contextual information around schooling was an essential part of any comparative investigation of what students know or have learned. In the series of meetings held in Hamburg and elsewhere that led up to the Pilot Twelve-Country Study in 1960, the goal among those gathered was to collect for the first time information on students' cognitive development in different national education systems (Husén, 1979). The theoretical framework for the Pilot Study was a general input-output model with student competencies in six subjects as the outputs and key aspects of schooling such as money, teacher competence, and teaching time as inputs.

Knowing that both the inputs and the outputs were embedded in different educational systems that pursued different curricula that reflected different national goals, the team planned to collect important curricula data. One experiment within the pilot was to obtain information on students' opportunities to learn what the test covered. The rationale for this was provided by Walker (1962):

> To establish the relevance of test items to pupils' learning opportunities is important both from the point of view of measuring achievement and from that of maintaining the goodwill of teachers whose pupils undergo the tests ... When the same test, or series of tests, is administered to pupils of different countries with different educational systems, it is unlikely that the items will be equally acceptable or equally useful in all of the countries concerned. The present inquiry was intended ... to estimate, if possible, the contributions made by stress on the topics in the curriculum ... to the accuracy with which pupils answered the questions (p. 63).

Of all the possible input variables considered – only a fraction of which they could include in the study – they viewed those from the pedagogical domain as most critical (Husén, 1979). As a pilot it was not possible to conduct all aspects of the study in every country. Nonetheless, the quote from Walker comes from the introduction of the chapter reporting on the success in gathering information on students' learning opportunities from teachers. In the later studies conducted by the IEA, reports always sought to draw attention to differences in school curricula in an effort to forestall individuals from "comparing the incomparable" by merely examining relative student performance (Husén, 1967a, 1983).

Thus the modern era of research in international comparative education that includes standardized measures of student performance in various subject matter areas of the school curriculum began with the 1960 Pilot Twelve-Country Study. The work around this first study subsequently led to the formation of the International Association for the Evaluation of Educational Achievement (IEA) (Husén, 1979; Keeves, 2011). Up until this pilot research project, comparative education largely consisted of rich descriptions of the education systems in the countries under consideration (Husén, 1967a; Keeves, 2011). Indeed, this assertion is easily verified through an examination of the table of contents for books on comparative education published prior to 1960, as well as many published afterwards.

For example, Hans' *Comparative Education* (1949) lists chapters addressing four major factors: natural factors such as race and language; religious factors; secular factors that reflect perhaps the zeitgeist of the education system such as humanism or socialism; and education in four democracies, a section that situates national education systems in their larger political-social-government regulatory contexts. Curiously, and to the point of our current consideration of students' opportunities to learn (OTL), there is little if any discussion of the actual content of the curricula offered by the national systems considered.

This lack of attention to the very core of the education enterprise – the specific subject matter content and topics that students study and learn – in the rich descriptions of education systems in the comparative education literature prior to the Pilot Study brings into sharp focus the revolutionary approach to such comparisons by the Pilot Study. The general input-output model that guided the enterprise together with the interest in using reliable and valid measures for all variables may well reflect their disciplinary backgrounds, with emphasis on learning and mental measurement rather than sociological, anthropological, or other perspectives on education. This emphasis on measuring the core of the educational enterprise may well be missed as the press and even many educators rush to view the latest results of what many now refer to as the "cognitive Olympics" (Husén, 1974).

The interest in the curriculum students studied came into sharper focus with the advent of the First International Mathematics Study (FIMS). Plans for conducting FIMS began as work on the Pilot Study was being completed and it became clear to all involved that such a study was indeed feasible. One of the central FIMS research questions was "what mathematics is taught, and in what order, country by country?" (Husén, 1967a) Mathematics had been chosen as the single subject for the first full blown international investigation, yet mathematics, even circumscribed by what is taught in school, encompasses a very large content domain. Consequently, this question of what was taught was compelling to the research team since "no single secondary school offers all that could be included in the subdivisions of mathematics usually taught in secondary schools collectively. They cannot. They must, therefore, choose what to include and what to exclude or ignore" (Husén, 1967a, p. 71).

Representatives of each country participating in FIMS consulted their national mathematics curriculum to propose topics suitable for an international assessment. The FIMS mathematics test covered topics that all participating countries included in their curricula as well as a few topics that were considered important but were covered only by a few of the countries (Husén, 1967a). These curriculum investigations served two purposes that became more explicit in subsequent IEA studies. First, the survey of the mathematics curricula served as a validity check for the items included on the mathematics assessment and, second, it served as an indicator of students' OTL with respect to the mathematics topics assessed. Actually, these two purposes are merely two sides of the same issue: the first having to do with the appropriateness of administering the test in a particular context; the second having to do with the appropriateness of interpreting students' performance in that context.

The FIMS OTL measure reflected the understanding that the test items are likely only part of any national curriculum. Teachers of the students that participated in the FIMS assessment were subsequently supplied with all the test items and asked to indicate if all or most, some, or few/none of the students in the group assessed had had the "opportunity to learn this type of problem" (Husén, 1967a, p. 167). The OTL variable used in analyses was a composite that indicated the extent to which the assessed students had had any learning opportunity for all the items on the test. This single OTL variable was highly correlated with scores both across and within countries (Husén, 1967b).

A similar OTL indicator was employed in the IEA's First International Science Study (FISS) (1966–1973) and was, again, found to be strongly associated with student performance. A different OTL indicator measured the extent to which the various science disciplines, e.g. biology, chemistry, earth science, and physics, were emphasized by a country's curriculum. This OTL emphasis measure was found to account "to a substantial extent" for the differences in student performance observed across countries (Keeves, 2011, p. 232). Given the recognized importance of curriculum in the studies up to this point, in a paper prepared for a meeting planning the second mathematics study (SIMS), David Walker asserted that "the second study should be approached as a curriculum survey. If the study focuses on the content of what is being taught, the relative importance given to various aspects of mathematics, and the student achievement relative to these priorities and content, then the international and national results can help in our understanding of comparable curriculum issues" (Travers and Westbury, 1989, p. 5). Indeed, the title of the first volume reporting on SIMS evidences the adoption of Walker's proposed emphasis for the study: *The IEA Study of Mathematics I: Analysis of Mathematics Curricula*.

Development of a Curriculum Model

The research and analytic model adopted in the Pilot Twelve-Country Study and the other early IEA studies has been referred to as a general input-output model (Husén, 1979). However, this does not answer the question of what it is a model of: student achievement; curriculum; or classroom instruction? An argument could be advanced for each and all of these. In the foreword to the Pilot Twelve-Country Study Report, Saul B. Robinsohn noted the success of the study to "supply what Anderson has lamented as 'the major missing link in comparative education,' which in his view is crippled especially by the scarcity of information about the outcomes or products of educational systems" (Foshay, Thorndike, Hotyat, Pidgeon, & Walker, 1962, p. 5). This had been the goal of the Pilot Study: an attempt to measure "intellectual functioning" in multiple national educational systems, precisely what so many viewed as lacking in comparative international educational research (Purves, 1987). Consequently, in the brief sixty-eight pages of the report on the Pilot Study, the words "assess" or "assessment" appear four times and the word "achievement" occurs forty-seven times.

Nonetheless, the goal of measuring "intellectual functioning" was not a goal in and of itself; it was a goal to add an aspect to comparative education with an eye on the central focus of education, i.e. the curriculum, evidenced in references to curriculum some thirty-two times in the same brief sixty-eight-page Pilot Study Report. The student achievement measure – constructed by psychologists with expertise in mental measurement – was not primarily of interest as an indicator of individual mental functioning. Rather, as Foshay stated in his portion of the report explaining the rationale for focusing on a student population at a point in which compulsory education ended, student measures were of interest "as representing the outcome of the educational system as a whole, rather than catching a student in mid-career, before the curriculum had been completed" (Foshay, 1962, p. 7).

Not all aspects of the Pilot Study were carried out in all of the participating countries. Walker (1962) reported on an exploration carried out in Scotland of students' "opportunity to learn." The success of using the student assessment items to obtain a sense of the degree to which students had had the opportunity to learn (OTL) the content required to respond successfully to the item was an important and transformative aspect of the Pilot Study (Purves, 1987). The study of intellectual functioning began to be viewed through a curricular lens with Walker's introduction of the "opportunity to learn" (OTL) construct. The implicit curriculum model and the OTL construct became a more explicit focus in subsequent IEA studies. The implicit model was formalized and presented in the report on the Second International Mathematics Study (SIMS) (Travers & Westbury, 1989). The SIMS Tripartite Curriculum Model, presented in Figure 2.1 (see Chapter 2), specifies three curricular components: the *Intended*, (standards, aims, goals, and expectations); the *Implemented* (the practices, content, activities, and strategies embedded in classroom instruction); and the *Attained* (the knowledge, ideas, and attitudes students develop from their curricular instruction).

In studies subsequent to the Pilot, researchers in participating countries with knowledge about the specific disciplinary focus of the proposed study would review their curricular documents – the *Intended* curriculum – to identify appropriate topics for inclusion on the student assessments. Teacher questionnaires were constructed in which they were asked to review each item and rate to what extent students in their classroom or school had had the opportunity to learn the "knowledge and skills required by the question" (Walker, 1962; Husén, 1967a, 1967b; Keeves, 1974; Travers & Westbury, 1989). These various OTL queries sought to measure the curriculum as *Implemented* in classrooms or, at least, if the narrowly defined assessment-relevant curriculum had been covered in the classroom.

In the science studies, the focus on assessment items for obtaining OTL measures shifted to the four fields – biology, chemistry, earth science, and physics – that comprise the majority of the science curriculum in most countries. For both the first science study, conducted as part of the Six Subject Survey that followed FIMS, and the Second International Science Study (SISS), a total of fifty-seven topics across these four fields were identified and presented to experts in each participating country to provide an indication of the degree to which each science field was emphasized in their students' curriculum. This rudimentary curriculum analysis was used to identify topics that would be appropriate to include on the student assessments (Keeves, 1992).

From these data, three validity indices for the student assessments were developed. One was a curriculum relevance index that measured "the extent to which the test covered the national curriculum" (IEA, 1988, p. 21). A second was a test relevance index that measured "the extent to which the items in the test were appropriate to the national curriculum" (IEA, 1988, p. 21). The third index, titled "the curriculum coverage index," provided an indication of the extent to which each country's curriculum addressed all the topic areas (curriculum) covered by all countries.

As described in Chapter 2, the 1995 Third International Mathematics and Science Study (TIMSS-95) substantially expanded the investigation of the intended curriculum through a rigorous and detailed analysis of national curriculum documents, i.e. curriculum guides or curriculum standards. TIMSS-95 also added to the Tripartite Curriculum Model by including textbooks (see Figure 2.2 – Chapter 2). Textbooks were conceived as potentially bridging the intentions of the *Intended* to what teachers did in their classrooms (Schmidt, McKnight, Valverde, Houang, & Wiley, 1997; Schmidt, Raizen, Britton, Bianchi, & Wolfe, 1997). TIMSS-95 and this aspect are more fully developed in the next chapter.

The Relationship between Time and OTL

Benjamin Bloom, probably best known for the taxonomy of learning objectives that bears his name, was one of the founding fathers that planned the Pilot Study and the IEA's first follow-up study, FIMS (Bloom, 1968; Foshay, 1962; Husén, 1967a, 1967b; Krathwohl, Bloom, & Masia, 1956). He is also well known for his Model of Mastery Learning, in which he posits that given sufficient time, all students are able to obtain a criterion level of understanding and learning (Bloom, 1968). As the E. L. Thorndike Career Achievement Award recipient in 1973, he opened his address with the statement, "All learning, whether done in school or elsewhere, requires time" (Bloom, 1974, p. 682).

The relationship between time and learning was implicit in the writings of William James around the turn of the last century. His "laws of learning," such as the laws of habit and repetition, refer to aspects of time in order to be efficacious (James, 1899). In addition, concepts such as effort and attention also include an implicit awareness of time as a required resource and a resource that requires appropriate management toward the goal of learning. This incipient recognition of the role time has in student learning was operationalized in the early IEA studies by the OTL items that explored the extent to which the curriculum – both the intended and implemented – covered (spent time on) the subject matter content addressed by the assessment items.

The important role of time in learning, however, was not explicit until Carroll's Model of School Learning (Carroll, 1963). Carroll had been involved in the early discussions that led to the Pilot Study and, as an experienced researcher in second language learning, he led the portion of the IEA's Six Subject Survey that investigated the learning of French as a second language (Carroll, 1975; Walker, 1976). In his analysis of these data, the amount of time spent studying the subject did not explain within-country differences in student performance as there was little variance in time within each country. At the country level, however, aggregated country means for time and student performance lined up almost perfectly (Carroll, 1975). His Model of School Learning grew out of his effort to identify the variables involved in successful second language learning and subsequently was profitably used in studies of other school subjects.

Carroll's model was geared toward providing a model that included all variables that impact student achievement and the interrelationship between those variables. His model includes three variables expressed as a function of time: aptitude, perseverance, and opportunity. The first two are student characteristics. Student *aptitude* is defined as the amount of time the student needs to learn the specific content. The amount of time a student is willing to work on learning is labeled *perseverance* and the amount of time available for the student to learn is labeled *opportunity to learn*. The two additional variables are not operationalized as a function of time. One has to do with the *quality of instruction* provided to the student and the other has to do with the student's *ability to understand* the instruction. The model is fully represented in Figure 6.1 and states that learning is a function of the time engaged in learning content divided by the time required to learn the content.

The OTL measures in the early IEA studies did not refer directly to "time" but whether topics were covered or stressed (that might indicate

$$\text{degree of learning} = f \left[\frac{\left(\begin{array}{c}\text{OTL or}\\ \text{Time allocated for learning}\end{array}\right) \times \left(\begin{array}{c}\text{Perserverance or percentage of}\\ \text{time actually spent engaged in}\\ \text{learning}\end{array}\right)}{\left(\begin{array}{c}\text{Aptitude or}\\ \text{Time actually needed to learn}\end{array}\right) \times \left(\begin{array}{c}\text{Quality of}\\ \text{instruction}\end{array}\right) \times \left(\begin{array}{c}\text{Ability to understand}\\ \text{instruction}\end{array}\right)} \right]$$

Figure 6.1   Carroll's Model of School Learning
Note. Reprinted by permission of the publisher. From Miriam Ben-Peretz and Rainer Bromme, *The Nature of Time in Schools: Theoretical Concepts, Practitioner Perceptions*, p. 17, New York: Teachers College Press. Copyright © 1990 by Teachers College, Columbia University. All rights reserved.

time) and alternatively what percentage of students would have received the opportunity. Note that those are two different concepts. The inference of "time" was by the analysts. OTL, in these studies, emerged as an important contextual variable for the observed variation in student performance. Carroll's model and the success of time in explaining student performance variation ignited several streams of research. Bloom's Model of Mastery Learning was developed from Carroll's model but highlighted the *aptitude* concept and set forth the idea that all students were capable of achieving a learning criterion if provided with the time each student needs to learn (Bloom, 1968, 1974). Others, dissatisfied with the mere provision of time, began to emphasize aspects or qualities of time that were thought to be more directly related to student learning. Thus in the Beginning Teacher Evaluation Study, in order to add more instructional meaning to the OTL concept measured, researchers investigated student engagement time – the time students actually spent on specific tasks and content along with a rating of the difficulty of the tasks they worked on – in addition to the allocated time (OTL) provided by teachers (Borg, 1980).

What is it about time that is important to student learning? How one approaches this question determines the nature of the variables included in models, research instruments, and analyses of student learning or performance. Opportunity can be considered in a binary manner; either students have or have not had experiences provided to them to learn specific content. Allocated time provides an indication of the extent to which students have had sufficient or appropriate time to learn. Yet some students make appropriate use of the time allocated for learning, while others may not. This led some to focus more on engaged time or academic learning time defined by an individual student's use of the time available for learning.

As Floden (2002) notes, each refinement in the notion of the sort of time most important for student learning came with an increase in the effort needed to identify, measure, and obtain the appropriate data. As obvious as Bloom's statement about time and learning seems now, it is also rather obvious that the simple provision of time is a necessary but not sufficient condition for student learning. As Carroll (1989) noted, the really important aspect of time is what happens during the time that is allocated for the learning task. This is referenced in his model by the *quality of instruction* variable and is something he concluded was probably unmeasurable.

Although the early IEA studies included OTL measures with a very small indication of emphasis (time), these measures were included in analyses more as indicators of content coverage of specific topics rather than measures of allocated instructional time. The two concepts are related but not identical and Carroll (1984) viewed the shift of focus to content coverage as a "valid and highly useful extension" of his model (p. 31).

OTL Measurement in Early IEA International Assessments

The title Carroll adopted for his model, a Model of School Learning, may seem to be a bit misleading as it is essentially a model that includes variables to explain the learning of an individual student. Perhaps a more precise title would be a Model for Student Learning in Schools and other Formal Settings. In the classroom settings, the OTL and *quality of instruction* variables would be typically held constant across students so the operative variables explaining individual student performance within the classroom would be those variables related specifically to the individual student, i.e. aptitude, perseverance, and ability to understand. Although the IEA studies sought to measure student intellectual functioning, the goal in each study was not to explain differences in performance at the individual student level but to use the individual student measurements as indicators of performance at different levels of the education system – countries, schools, or classrooms – to which the students belonged.

In other words, the goal in obtaining measures of students' learning opportunities was to provide context for aggregated student performance rather than that of the individual, within-classroom student performance. Given this IEA research goal it follows that the sampling of classrooms and the gathering of OTL data from the teachers of the tested students is an appropriate sampling design. A better design statistically might be to sample a small number of students from each classroom and obtain the teacher and OTL data from all the classrooms in the school. The best method from a sampling point of view would be to take all classrooms and all students.

Table 6.1 provides a summary of the various ways the contextual OTL information was obtained in IEA studies conducted in the 1960s and 1970s. Context OTL indicators about the intended curriculum were provided by mathematics experts. Teachers provided OTL indicators about the implemented curriculum. SISS was the only study that assembled teachers in the sampled schools and had them provide OTL information as a group. Otherwise, a single teacher provided OTL information that applied to all students in that teacher's classroom. To the extent that most teachers within a country provided the same response, the lack of differences in teachers' responses could not possibly explain the variation in students' responses within any one country. This was offered as a possible explanation for the lack of significant relationships in the curriculum-related regressions conducted with the Pilot Study OTL data (Walker, 1962). In FIMS, very low within-country correlations were observed between the teacher-provided OTL and student performance, which likely reflected the homogeneity of OTL in those countries or reflected that the OTL measures themselves were not at a specific enough level in terms of the content topics included. It is also possible that the nature of the OTL items lacked validity.

Correlations across countries, however, were generally greater than .6 for the three general FIMS populations [.64, .73, .80] (Husén, 1967b, p. 170). Husén noted that in the cross-country regression analyses with teacher variables, nearly all the student performance variance was accounted for by the opportunity measure (Husén, 1967b, p. 272).

In an attempt to obtain OTL data that would improve the relationship of OTL to performance at the within-country level, SIMS asked students to respond to test items in a manner similar to their teachers. This was also viewed as a validity check on teachers' OTL judgments. For each item on their test, students were asked whether they had been taught the mathematics required for that item during their current academic year, in a year prior to the current one, or never (Robitaille & Garden, 1989).

Analyses of these data from a few selected countries, however, did not yield easily interpretable results and cast considerable doubt on the usefulness of these OTL data. Not all countries chose to include this additional task as part of the student assessment. Some were skeptical about the usefulness of student judgments; some were concerned about the additional response burden on students; while some shared both of these concerns. Nonetheless, most countries did include this additional student OTL in their assessments but the United States elected to collect these data only for the forty-item core test.

Table 6.1 *OTL measurement in IEA studies prior to the 1995 TIMSS*

Focus	Prompt	Question	Reference

Exposure/ Coverage	Test Items	Teachers in tested schools were given all assessment booklets and asked to rate each item twice.
		 1. Consider the degree to which the knowledge and skills required by the question are typically covered in the instruction of pupils in classes like yours. Give a rating on the following three-point scale.
		   Rating 1. Stressed: well covered in class and in homework (if any).
		   Rating 2. Included but not stressed; touched on but not dealt with extensively, intensively, or repeatedly.
			Pilot
			 Walker, 1962, page 63

Teachers		Rating 3. Not included.
		 2. Consider next the extent to which pupils have opportunity in the home and in the community to use skills or encounter knowledge such as those involved in the test question. Decide whether there is **considerable**, **some**, or **little** exposure to such experiences. Use the following rating scale.
		   Rating A. Considerable exposure
		   Rating B. Some exposure
		   Rating C. Little or no exposure	

Exposure/ Coverage National mathematics expert panel	Test Items	Each country's National mathematics experts panel rated each item twice:
		 1. Item Exposure via Assessments
		   A. the problem or task is almost identical to problems or tasks likely to appear on the terminal examination paper for at least 76 percent of all students taking this set of tests.
		   B. the problem or task is almost identical to problems or tasks likely to appear on the terminal examination paper for between 25 percent and 75 percent of all students taking this set of tests.
		   C. the problem or task is almost identical to problems or tasks likely to appear on the terminal examination paper for less than 25 percent of all students taking this set of tests.
		 2. Item Exposure via the Curriculum
		   A. this item would be in the curriculum for at least 76 percent of all students taking this set of tests.
			FIMS
			 Husén, 1967a, pages 136–147

Table 6.1 (cont.)

Focus	Prompt	Question	Reference
		B. this item would be in the curriculum for between 25 percent and 75 percent of all students taking this set of tests.	
		C. this item would be in the curriculum for less than 25 percent of all students taking this set of tests.	
Exposure/ Coverage Teachers	Test Items	Teachers were provided all the test items and asked to "examine each question in turn and indicate in the way described below, whether in your opinion A. All or most (at least 75%) of this group of students have had an opportunity to learn this type of problem. B. Some (25% to 75%) of this group of students have had an opportunity to learn this type of problem. C. Few or none (under 25%) of this group of students have had an opportunity to learn this type of problem."	**FIMS** Husén, 1967b, pages 167–168
Exposure/ Coverage Teacher Group	Test Items	*In each school, teachers as a group* were asked to rank each item on a 5-point scale: • All students (4) • More than 75% of students (3) • Between 25% & 75% of students (2) • Less than 25% of students (1) • None of the students (0)	**SISS** Keeves, 1974
Exposure/ Coverage Teacher Group	Test Items	*In each school, teachers as a group* asked to "indicate the grade level at which the concepts tested by the item would usually be first taught in your school: • Grade 4 or earlier • Grade 5 • Grade 6 • Not at all"	**SISS** (Burstein notebook, 1991)
Exposure/ Coverage Teachers	Test Items	Teachers of each sampled student classroom asked 3 questions about each test item: 1. What percentage of the students from the target class do you estimate will get the item correct without guessing? a. Virtually none b. 6–40% c. 41–60%	**SIMS** (Burstein notebook, 1991)

Table 6.1 (cont.)

Focus	Prompt	Question	Reference

		d. 61–94%
		e. Virtually all
		2. During this school year, did you teach or review the mathematics needed to answer the item correctly? [yes; no]
		3. If, in this year, you did NOT teach or review the mathematics needed to answer this item correctly, was it because,
		  a. It has been taught prior to this school year?
		  b. It will be taught later (this year or later)?
		  c. It is not on the school curriculum at all?
		  d. For other reasons.
			(Travers & Westbury, 1989 for parts 2 & 3)

Coverage Teachers	Topics	The topics given below may be included in your instructional program. Circle the appropriate response code to show whether for students in the target class the topic was: a. Taught as new content. b. Reviewed and then extended. c. Reviewed only. d. Assumed as prerequisite knowledge and neither taught nor reviewed. e. Not taught and not assumed as prerequisite knowledge. Examples of topics listed: 1. The concept of positive and negative integers. 2. Solving linear equations. Ex. Solve $4x - 3 = 19$.	**SIMS** (Burstein notebook, 1991)
Coverage Teachers	Instruction Activities	Indicate the amount of time spent on each of the following activities (that is, demonstrations, explanations, students doing computations, using manipulative, etc.) with your target class. Circle the estimated number of class periods. If more than 10 periods were spent on any topic, specify the number of periods on the blank. Examples of activities listed: 1. Activities related to determining precision, accuracy, percent error and relative error. 0 1 2 3 4 5 6 7 8 9 10 ____ 2. Activities related to the concept of π. 0 1 2 3 4 5 6 7 8 9 10 ____ 3. Activities related to finding the area of circles. 0 1 2 3 4 5 6 7 8 9 10 ____	**SIMS** (Burstein notebook, 1991)

Table 6.1 (*cont.*)

Focus	Prompt	Question	Reference

Emphasis in Instruction
 	
Teachers
	Problem Types	Several types of problems are listed below which may have been included in your instructional program. Circle the appropriate response code to indicate the degree to which a particular type of problem was studied by the target class:
		a. Emphasized (used as a primary type of problem, used extensively or frequently).
		b. Used, but not emphasized.
		c. Not used.
		Examples of problems listed:
		1. Distance-Rate-Time problems.
		   How long does it take a rainstorm to travel 360 km at a rate of 45 km per hour?
		2. Area-Volume problems.
		   The Great Pyramid in Egypt has a square base measuring 240 m on a side. Its altitude in 160 m. What is its volume?
			SIMS (Burstein notebook, 1991)

Up to this point we have considered OTL strictly according to the Carroll model, with its emphasis on the time associated with specific content. The incorporation of OTL in studies of student learning and achievement provided important contextual information about student experiences and began to open up the "black box" of what occurs in schools and classrooms (Darling-Hammond & Snyder, 1992). A greater examination and exploration of the "black box" was the explicit goal of SIMS as it was described as a curriculum study. Thus the teacher questionnaire – the primary means to explore the implemented curriculum – included many different types of questions exploring the extent of student exposure to specific instructional topics, activities, and types of problems.

Examples of these that are not available in any of the official SIMS reports are included in Table 6.1. The full set of questionnaires and the data are available from the international databank now in Sweden (https://ips.gu.se/english/research/research_databases/compeat/Before_1995/SIMS). The inclusion of these types of questions begins to move from the Carroll concept of OTL (time spent on specific instructional content) into an attempt to capture some aspect of Carroll's *quality of instruction* concept. In the activity leading up to the 1995 TIMSS, Leigh Burstein and his colleagues were tasked with gathering items from previous international and other research studies that had been used to explore OTL. In his memo introducing the large resultant notebook, he observed that the "lines of demarcation are still fuzzy regarding where OTL ends and classroom processes begin" (Burstein, June 19, 1991). An examination of the SIMS example items assembled by Burstein confirms this (see Table 6.1). The last two items in the table regarding activities and topics of instruction, while having a time dimension, may well be viewed as crossing into the instructional practice (quality) domain.

The TIMSS-95 study using the Burstein memo built on the SIMS OTL measures. The TIMSS-95 measures are described in greater detail in the next chapter, as it took the measurement of opportunity to learn to a new level as a curriculum study in and of itself. Not only did TIMSS-95 include OTL in the teacher questionnaire as a measure of the *Implemented* curriculum to be related to student performance but it included a major, in-depth investigation of the *Intended* and *Potentially Implemented* curriculum as a separate study in conjunction with the main student assessment investigation (Schmidt, Jorde, et al., 1996: Schmidt, McKnight, Valverde, Houang, & Wiley, 1997; Schmidt, Raizen, Britton, Bianchi, & Wolfe, 1997). Many of the original curriculum study publications focused solely on characterizing the curricula around the world, describing in detail their many similarities and differences (Schmidt,

McKnight, Valverde, Houang, & Wiley, 1997; Schmidt, Raizen, Britton, Bianchi, & Wolfe, 1997; Schmidt, McKnight, & Raizen, 1997). The derivative TRENDS studies and their measures of OTL are also covered in Chapter 7.

Table 6.1 indicates that for several IEA studies, national expert panels rated items on an appropriateness scale according to official curriculum documents. This served both as a validity check for the student assessment in that country as well as an indication of the intended curriculum. However, this intended curriculum measure only represented the intended curriculum in so far as what was also on the test; what had occurred before as well as what else may be included in the curriculum for that grade was ignored. The implemented curriculum was similarly measured, focusing only on the topics assessed and ignoring anything else that may have occurred during the year in which students were assessed. During the research that led up to the 1995 TIMSS, interest in the curriculum was broadened in recognition that the assessed curriculum represented significantly different proportions of the intended and implemented curriculum in different countries. The Model of Potential Educational Experiences (see Figure 2.3) that was developed to guide the 1995 TIMSS embedded this broadened concept of curriculum (Schmidt, Jorde, et al., 1996).

OTL Measurement in the PISA 2012 International Assessment

In the 2012 study, PISA, for the first time, included the collection of OTL data related to mathematics. What made this initial effort different from the traditional IEA studies was driven by the PISA sampling design: Students were randomly sampled within schools without regard to classrooms and, consequently, there was no teacher data collection. This left the task of providing OTL data to the students. Using students as the primary and only informant on OTL had not been done in IEA studies. However, similar in some ways to the idea in the IEA studies of giving teachers or country representatives a list of topics to which they would respond as to whether the topics were covered in the country's curriculum, students were given a list of mathematics content and asked to indicate the degree to which they were familiar with each.

Focusing the list on the topics they had covered in their current grade would have been very difficult, if not impossible, to accomplish since even within countries, fifteen-year-old students were in different grades in all but two countries. It would also not have been appropriate given the cross-sectional nature of the PISA sample and the fact that performance on a literacy assessment develops through the cumulative exposure to the curriculum and not at a specific grade.

The fifteen-year-old students typically had eight to eleven years of schooling in mathematics, depending on the country. A complete cumulative list would have been too long, as some students would have had eight years of schooling, while others would have had eleven, making the list of content areas burdensome and excessively long.

This led to a marriage of the methodologies of both PISA-2012 and TIMSS-95. Using the TIMSS-95 index of grade placement (IGP - described more fully in Chapter 7), a set of topics were chosen to represent the full range of mathematics across all twelve grades. Given the hierarchical nature of mathematics as represented by the index, student responses to the list would enable an estimate of how far the student had progressed in their mathematics studies. The students were asked to specify how familiar they were with each of the thirteen selected topics. Those chosen in terms of typical coverage spanned grades 3 and 12, e.g. ranging from whole number division to complex numbers.

Student responses were scaled from 0 to 4, with 0 representing the category, "never heard of it." Summing across all the IGP weighted concepts (using the typical grade, as defined by the IGP, at which the topic occurred as the weight) provided an estimate of the cumulative rigor or the complexity of the content of the mathematics curriculum that students had experienced in school up to and including the age of fifteen (see Appendix C) (Achieve, 2004; Cogan, Schmidt, & Wiley, 2001; Schmidt & Cogan, 2009; Schmidt, Cogan, & McKnight, 2010; Schmidt, Cogan, Houang, & McKnight, 2011; Schmidt & McKnight, 2012).

The resulting measure of OTL is multi-faceted, being based on three distinct aspects of OTL: (1) the coverage of the mathematics content itself, (2) the degree of familiarity with each topic (a surrogate for exposure), and (3) the content complexity, estimated from the TIMSS-95 IGP. The variable was finally standardized between 0 (no familiarity with any of the thirteen topics listed) and 3 (Schmidt, Cogan, & Guo, in press).

This measure of school or formal mathematics OTL was found to be statistically significantly related to literacy performance at the within-school level as defined by the total score in each of the sixty-plus participating countries (Schmidt, Cogan, & Guo, in press). The same significant relationship was found for virtually all countries for the seven PISA sub-scores. The results presented in Table 6.2 are from a two-level hierarchical analysis defined at the between- and within-school levels.

Table 6.2 *Average significant estimates and percentage of PISA countries with statistically significant relationships of OTL to PISA performance for each of the three measures of OTL*

	Effect	Literacy	Change	Quantity	Space	Data	Employ	Formulate	Interpret	Problem solving
Within-School Level	Grade	22.7 / 94%	26.3 / 91%	24.5 / 91%	20.5 / 89%	23.5 / 91%	22.9 / 95%	23.6 / 91%	23.9 / 92%	23.6 / 93%
	SES	10.9 / 95%	11.3 / 29%	11.4 / 32%	11.4 / 29%	10.9 / 32%	10.9 / 30%	11.6 / 32%	11.0 / 29%	9.7 / 31%
	Applied Math	12.3 / 35%	14.5 / 29%	13.9 / 32%	13.5 / 29%	15.2 / 32%	13.6 / 30%	14.2 / 32%	14.8 / 29%	12.4 / 31%
	Appl. Math Quadratic	-4.9 / 48%	-5.4 / 36%	-5.6 / 39%	-5.0 / 33%	-5.2 / 48%	-4.8 / 44%	-5.1 / 44%	-5.2 / 50%	-5.2 / 40%
	Word Problems	4.5 / 71%	5.2 / 59%	5.2 / 65%	4.7 / 59%	4.5 / 64%	4.2 / 73%	4.5 / 65%	4.6 / 59%	3.7 / 69%
	School Math	43.0 / 100%	46.4 / 100%	43.3 / 100%	45.2 / 100%	39.4 / 100%	42.3 / 100%	46.4 / 100%	41.1 / 100%	38.7 / 98%
Between-School Level	SES	40.5 / 98%	42.7 / 97%	43.1 / 98%	41.1 / 98%	39.5 / 98%	39.0 / 97%	43.4 / 98%	42.2 / 97%	40.9 / 98%
	Applied Math	112.2 / 58%	119.5 / 61%	133.0 / 67%	148.2 / 67%	109.0 / 61%	113.8 / 61%	116.2 / 62%	102.0 / 71%	160.3 / 76%
	Appl. Math Quadratic	-31.3 / 62%	-36.5 / 67%	-35.8 / 64%	-33.7 / 71%	-34.2 / 73%	-31.8 / 65%	-32.4 / 59%	-31.4 / 71%	-47.2 / 86%
	Word Problems	16.4 / 73%	21.0 / 79%	16.8 / 80%	19.5 / 67%	16.8 / 80%	16.2 / 76%	18.6 / 76%	17.8 / 74%	13.4 / 76%
	School Math	88.9 / 98%	97.3 / 98%	91.5 / 98%	92.6 / 98%	82.2 / 97%	90.5 / 98%	94.5 / 98%	85.5 / 97%	78.6 / 98%

Table 6.2 (*cont.*)

	Effect	Literacy	Change	Quantity	Space	Data	Employ	Formulate	Interpret	Problem solving
	Grade	100%	100%	100%	100%	100%	100%	100%	100%	100%
	SES	98%	98%	100%	100%	100%	98%	100%	97%	98%
	Applied Math	73%	71%	79%	77%	76%	73%	73%	80%	83%
	Appl. Math Quadratic	83%	77%	76%	83%	91%	79%	79%	89%	90%
Either Level	Word Problems	86%	88%	88%	79%	86%	91%	88%	88%	95%
	School Math	100%	100%	100%	100%	100%	100%	100%	100%	98%

Note. The table is derived in part from "The Role that Mathematics Plays in College- and Career-Readiness: Evidence from PISA," by W. H. Schmidt, L. S. Cogan, & S. Guo, 2018. Copyright 2018 by the Taylor & Francis Group, Informa Group Plc.

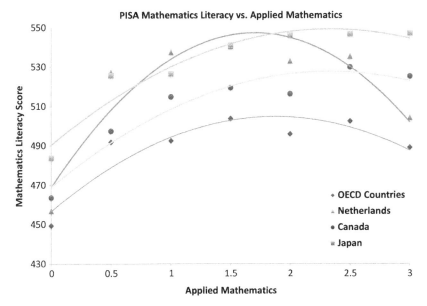

Figure 6.2 The quadratic relationship of Applied Mathematics to performance illustrated for four countries
Note. The figure is derived in part from "Excellence and Equality in Mathematics Education," by W. H. Schmidt, 2017. Copyright 2017 by the American Affairs Foundation Inc.

The PISA mathematical literacy focus led to the development of a set of OTL items to measure the exposure of students to the application of mathematics used to solve real-world problems. To do this, two additional categories of items were developed: traditional word problems as typically found in mathematics textbooks; and applied problems related both to mathematics – like finding the volume of a pyramid – and to real-life problems that involve quantities such as drawing conclusions from statistical graphs related to important societal issues or determining how much wood is needed to build a fence.

Two examples were developed for each of these two categories, and students were then asked on a four-point scale the degree to which they had been exposed to such types of problems in their classroom instruction and in the school-related tests they experienced. These formed two additional scales – word problems and applied mathematics. The two-level statistical results for these application-oriented measures of OTL are also summarized in Table 6.2. Unlike formal school mathematics, which was statistically significant in all countries, applied mathematics was significant in around 80 percent of the countries (either as a linear or quadratic relationship at either the within- or between-school level), as was word problems.

The relationship of applied mathematics OTL was statistically significantly related to literacy performance, but the nature of the relationship was typically not linear but quadratic. This suggests a positive relationship up to an inflection point after which the relationship was negative. Figure 6.2 illustrates this relationship for several countries.

PISA has just begun to incorporate OTL measurement as a regular part of its data collection in the mathematics-focused studies. The results from a humble beginning were quite convincing and important, adding a new dimension to PISA results (see chapter 3 in *PISA 2012 Results: What Students Know and Can Do: Student Performance in Mathematics, Reading and Science Volume I*).

CHAPTER 7

The 1995 TIMSS Curriculum Analysis and Beyond

The previous chapter described in detail the various ways in which the concept of opportunity to learn was measured both in PISA and in the IEA studies leading up to and including the second in the series of mathematics and science studies – SIMS and SISS. In this chapter we focus on the Third International Mathematics and Science Curriculum Study, but also include the TRENDS derivative studies.

The description of the measurement of OTL in SIMS captures the essence of how those previous four IEA studies viewed OTL essentially as a means to determine the appropriateness or fairness of the test. This idea was clearly noted in the First International Mathematics Study (FIMS) by Husén (1967a), "With some justification, one might paradoxically say that the tests devised for the IEA study are equally appropriate or inappropriate to all the countries participating in the study" (p. 21). In the same vein, the authors of the SIMS curriculum volume described this more fully:

> In common with earlier IEA Studies, the conceptualization of SIMS provides a means for determining the appropriateness, or fairness, of the tests that were used. This determination can be made at the level of the intended curriculum, i.e. to what extent is the mathematics content of the test (the SIMS pool of items) "intended to be covered" by the curriculum of the target population in each system? It can also be made at the level of the implemented or taught curriculum – to what extent has the mathematical content of the test actually been taught to the students who were tested? (Travers & Westbury, 1989, p. 15)

The TIMSS-95 study building on that work developed the most thorough, extensive, and scientifically rigorous system of OTL measurement ever accomplished in an international study. In fact, what emerged was a study in its own right, yet also a significant component of TIMSS-95 along with a video study of education in Japan, Germany, and the United States. Both of these were designed to provide a richer context for understanding the assessment results. As it developed, it moved beyond the traditional IEA goal of determining "the appropriateness or fairness of the test" to a full blown study of curriculum as an additional important outcome of educational systems – a policy-relevant variable in its own right.

In that vein, the curriculum study also included looking at the structure and decision-making policies associated with the curriculum, as well as providing a detailed description of the curriculum from a country's standards – including the content development of topics over the grades (intended curriculum) – specifying the content, pedagogical style, and approach taken in the country's textbooks (potentially implemented curriculum), and what content was actually taught by the teachers (implemented curriculum).

All of this was carried out for both mathematics and science over the K–12 school system for over forty countries – in effect creating the first extensive, integrated, and data-driven international curriculum study (Schmidt, McKnight, Cogan, Jakwerth, & Houang, 1999; Schmidt, McKnight, Houang, Wang, Wiley, Cogan, & Wolfe, 2001; Schmidt, McKnight Valverde, Houang, & Wiley, 1997; Schmidt, Raizen, Britton, Bianchi, & Wolfe, 1997). For this reason, the description of the TIMSS-95 approach to OTL measurement is appropriately presented in its own chapter. Also included is a description of the way key features of the TIMSS-95 curriculum analysis were included in the subsequent five TRENDS studies.

The core of the TIMSS-95 approach rested on the desire to develop a scientific approach to the collection and quantification of content coverage (OTL) so as to develop measures that characterize one of the central aspects of schooling – which involves the interaction of teachers with students around the learning of specific content, in this case mathematics and science content.

For TIMSS-95 this required the development of a framework not just to develop a test blueprint or to determine fairness, as was done in one way or another in all previous IEA studies, but to characterize the full extent of the K–12 curriculum in terms of content coverage, the degree of emphasis associated with each topic, and the sequencing of content over the grade levels. The resulting portraits allowed for comparisons of countries' curricula alongside comparisons of student performance. This builds on, but also moves beyond, traditional IEA OTL measurement focusing on tested-grade portraits of content coverage to a characterization of the whole of the curriculum (K–12), itself an important policy-malleable component of a country's education system. This also helps in the interpretation of cross-sectional data since performance at a given grade is likely related to the student's cumulative opportunities to learn, especially in mathematics, given its hierarchical nature.

Defining such a framework was the first task undertaken in the SMSO/TIMSS-95 study and that ended up taking almost a year to fully develop. To make a framework part of a formal measurement system of curriculum required the specification of the content domain over some population of grades and countries. Before TIMSS-95, this was done in somewhat of an ad hoc fashion, as the researchers in the participating countries often chose examples of the test items they wanted to study and then classified them, creating, in effect, the test blueprint or framework.

Another approach was to conduct an informal survey of countries as to what was taught at a particular grade level. Either way, a draft was then circulated to countries for modifications and input. For example, in SIMS, country representatives were asked to rate the test items according to their appropriateness with respect to their national curriculum or its equivalent (Travers & Westbury, 1989).

TIMSS-95 chose a new approach over that in SIMS because, in part, there were criticisms of the resulting product – a content-by-process matrix (Romberg & Zarinnia, 1987). The TIMSS-95 Curriculum Framework for Mathematics and Science (1993) summed up the expressed concerns as follows:

> ... use of a content-by-cognitive-behaviour grid fails to take into account the interrelatedness of content or of cognitive behaviours, and that this forces the description of information into unrealistically isolated segments. In addition, questions have been raised about the usefulness of such grids for characterizing curricula and of their flexibility in accommodating different theoretical notions about how students learn (Robitaille, Schmidt, Raizen, McKnight, Britton, & Nicol, 1993, p.42).

The TIMSS-95 principle used in specifying the domain was inclusivity, essentially using the union principle across the forty-plus participating countries: All topics covered in grades K–12 for all countries were included. What turned out to be helpful and, in the final analysis, ground breaking was that such an exhaustive detailing of the mathematics content covered in schools over all primary and secondary grades based on a large and diverse set of countries provided the richest and most detailed context for interpreting the variation in country performance. As ground breaking as this was, it also led to what may be the greatest contribution of TIMSS-95 – the quantification of one of the most fundamental aspects of schooling – content coverage. Having such a measure of schooling allowed quantitative analyses linking SES and OTL to performance. Relating a quantitative measurement of SES to performance already had a long tradition, which, without such a measure of schooling, often led to a misleading of the importance of SES in relation to student achievement without recognizing the relationship of SES to OTL (see Chapter 8).

Frameworks were developed to describe the K–12 domain for mathematics and for science education around the TIMSS-95 world. In both cases experts from the participating countries convened and developed the framework; it took over a year in order to achieve the needed cross-country consensus and to make sure the framework was indeed exhaustive for the union of the curricula of the forty-plus participating countries.

The two TIMSS-95 frameworks were developed to have a parallel structure. They both had a detailed specification of content but also two other dimensions – performance expectations and perspectives. Performance expectations were like the cognitive behavior dimensions included in previous IEA studies and dealt with areas such as reasoning and problem solving in mathematics, and abstracting and deducing scientific principles and gathering data in science. Perspectives focused on attitudes. However, countries' standards and textbooks yielded little insight into this dimension and consequently did not play much of a role in TIMSS-95. Figure 7.1 summarizes the structure of the two frameworks.

Perhaps the best summary of the role the frameworks played in TIMSS-95 is:

> The TIMSS curriculum frameworks were constructed to be powerful organizing tools, rich enough to make possible comparative analyses of curriculum and curriculum change in a wide variety of settings and from a variety of curricular perspectives. The frameworks had to allow for a given assessment item or proposed instructional activity to be categorized in its full complexity and not reduced to fit a simplistic classification scheme that distorted and impoverished the student experience embedded in the material classified (Robitaille, D. F., Schmidt, W. H., S. Raizen, C. McKnight, E. Britton, & C. Nicol, 1993. P. 42).

The content dimension of the two frameworks was also developed so as to allow measurement at different levels of specification using a nested design with a more expanded specification of broader topics at the lower levels of the hierarchy. This allowed for creating summarizations at different levels of specification and permitted reporting by aggregation to higher levels of the framework. For example, the mathematics framework includes a broad category of geometry – position, visualization, and shape – but also has more specific subdomain topics such as vectors and coordinate geometry.

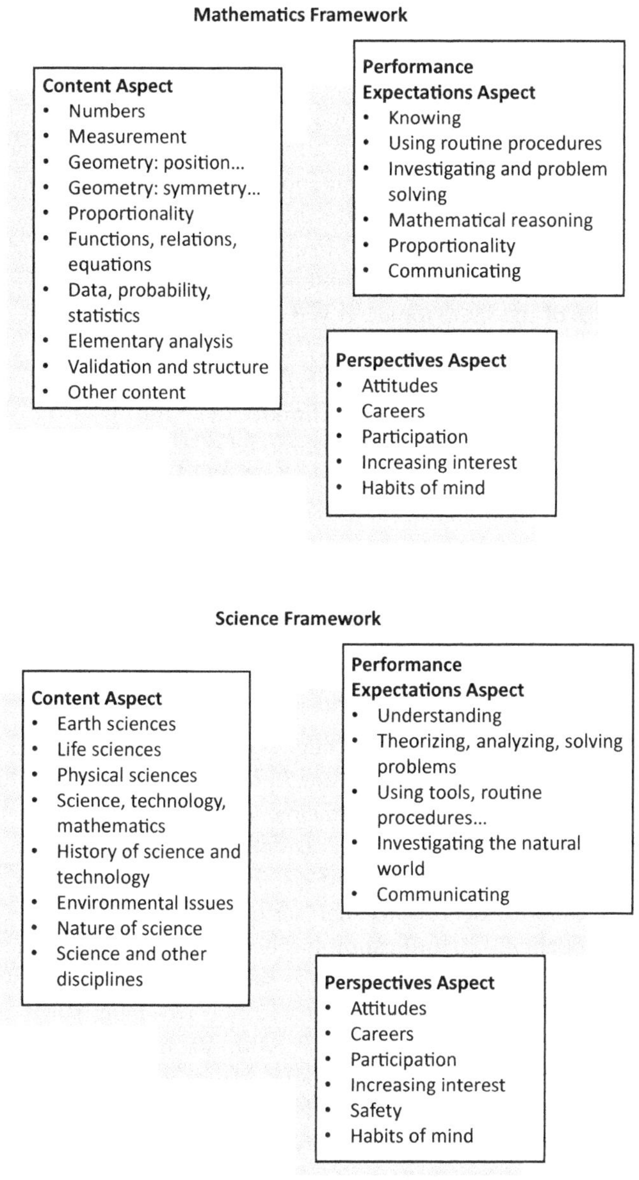

Mathematics Framework

Content Aspect
- Numbers
- Measurement
- Geometry: position...
- Geometry: symmetry...
- Proportionality
- Functions, relations, equations
- Data, probability, statistics
- Elementary analysis
- Validation and structure
- Other content

Performance Expectations Aspect
- Knowing
- Using routine procedures
- Investigating and problem solving
- Mathematical reasoning
- Proportionality
- Communicating

Perspectives Aspect
- Attitudes
- Careers
- Participation
- Increasing interest
- Habits of mind

Science Framework

Content Aspect
- Earth sciences
- Life sciences
- Physical sciences
- Science, technology, mathematics
- History of science and technology
- Environmental Issues
- Nature of science
- Science and other disciplines

Performance Expectations Aspect
- Understanding
- Theorizing, analyzing, solving problems
- Using tools, routine procedures...
- Investigating the natural world
- Communicating

Perspectives Aspect
- Attitudes
- Careers
- Participation
- Increasing interest
- Safety
- Habits of mind

Figure 7.1   The three aspects and major categories of the mathematics and science frameworks
Note. Adapted from "Curriculum frameworks for mathematics and science," by D. F. Robitaille, W. H. Schmidt, S. Raizen, C. McKnight, E. Britton, & C. Nicol, 1993, p. 46. Copyright 1993 by International Association for the Evaluation of Educational Achievement. Publisher: Pacific Educational Press.

In addition, the "grain size" (level of specificity of a topic) of the content dimension was not made to be equal but rather a representative list of topics and categories. Consequently, attention had to be paid to varying levels of specificity when interpreting results of the framework (Houang & Schmidt, 2008). It also points out a limitation of the study; some topics that are more specifically defined are more able to support detailed analyses than is the case for topics that are more generally defined. By allowing topics to include varying degrees of subcategories, the framework permits either a broad (macro) view of content or a specific (micro) view of the subcategories of the same topics. It was broad enough to include any topic included in any participating country's curriculum and precise enough to accurately portray curriculum complexity in a comparable manner (and in a manner that may be analyzed) (Schmidt et al., 2001).

The mathematics framework is composed of ten major categories, each of which contains different numbers of subcategories, for a total of forty-four content-specific topics, and twenty-one performance expectations. The science framework, on the other hand, is composed of eight major categories, with a total of seventy-nine content-specific topics, and twenty performance expectations (Schmidt et al., 2001). The international mathematics and science frameworks are found in Appendix A. They represent the domains over which the mapping functions were specified defining the measurements of OTL.

The important feature of these frameworks with respect to the quantification of OTL was that the topics listed provided a mutually exclusive and exhaustive perspective of the domains, at least with respect to K–12 mathematics and science education as defined in the forty-plus countries participating in the original TIMSS-95. Establishing these characteristics permits meaningful definitions that are quantifiable and can be used in statistical modeling. Quantification has been an important element in the advancement of other areas of science such as astronomy, in which the quantitative measurement of distance using red light shifts advanced astrophysics. More recently, biology, with its increased quantification related to the mapping of the genome, has experienced great scientific advances, pushing some to suggest that, while the twentieth century was about physics, the twenty-first century will belong to biology. We believe that this work, although less grandly, will help advance educational research by providing a quantitative measure of schooling, at least for the area of mathematics.

The founding fathers recognized the necessity of having measurements of OTL in order to understand the results of international assessments.

They knew that without such descriptions of what countries taught in mathematics, for example, international assessments would have little or no interpretive value as to the why and, in the worst case, could lead to bad and unwarranted conclusions that were actually harmful to education worldwide, such as believing countries do better because they have a special gene pool. The founding fathers' instincts were right, but they did not have the sophisticated tools to accomplish what they knew was needed.

This limitation is also somewhat true for education research more generally. There are many studies that examine the relationship of certain features of schooling to performance without taking into account differences in OTL. Following the original Coleman study, the debates raged and still do about whether it is schooling or family background that is most important to student achievement (Averch, Carroll, Donaldson, Kiesling, & Pincus, 1974; Baker, Goesling, & LeTendre, 2002; Bidwell & Kasarda, 1980; Bowles & Levin, 1968; Chudgar & Luschei, 2009; Gamoran, 1996; Gamoran & Long, 2007; Greenwald, Hedges, & Laine, 1996a; Hanushek & Kain, 1972; Heyneman & Loxley, 1983; Jencks, Smith, Acland, Bane, Cohen, Gintis, Heyns, & Michelson, 1972; Plowden, 1967a; Plowden, 1967b). The problem until more recently was that measures characterizing schooling were about broad related characteristics such as school organization, teacher characteristics, funding, etc., and not about the basics such as a measure of the actual content students were studying – a part of the "black box" in which school learning takes place involving students, a teacher, and the actual content being studied.

A related assumption of those studies, either explicitly or implicitly, was that at a given grade level within a country all students were studying the same mathematics content. More recently, we know this is not the case in the United States at least (Cogan, Schmidt, & Wiley, 2001; Schmidt & McKnight, 2012), and as a result of TIMSS-95 it is certainly not the case cross-nationally. The founding fathers instinctively knew this and, as a result, started the field of international educational assessment off on the right foot.

The absence of good measurements of opportunity to learn in the past likely resulted in the conclusion by many that SES was the major driving force in school learning rather than schooling itself. In many ways those of us who have worked on TIMSS-95 and PISA believe the greatest contribution these studies have made is the development of quantitative measures of OTL that are now a part of both TRENDS and PISA as well as national studies in the United States and other countries. The new US longitudinal study of middle school is a good example, as it focuses on describing what mathematics students are studying (Institute of Education Sciences, n.d.).

Chapter 2 described the TIMSS-95 model of curriculum, identifying four elements: the intended, the potentially implemented, the implemented, and the achieved (see Figure 2.2). For all four of these levels individuals or documents needed to be identified from which the needed information could be collected. The mathematics and science frameworks provided the common definitions of content from which all indicators of opportunity to learn (OTL) were to be developed. We now turn to each of the levels, first identifying the relevant sources that can be used to collect the data and then indicating what quantitative indicators can be developed with examples from TIMSS-95. We first consider the measurement of the intended curriculum.

The Intended Curriculum

Virtually every country in the world has some type of document, referred to by many different names, that defines for a given subject matter the topics that are expected to be learned at each grade in the educational system. Such specifications, often called standards or curriculum guidelines, express the expectation that all children (or some subset of them) at a given grade will cover the listed topics. These are policy documents specifying what the country wants students to know and at what grade. Across countries they vary as to the level of detail, ranging from a simple list of topics to a much more detailed elaboration of the topic as well as approaches to the teaching of the topic. Rarely, but sometimes, the document also indicates the amount of time that should be devoted to covering each topic.

The approach used in TIMSS-95 was that each standard in the country document was mapped (coded) onto the appropriate mathematics or science framework. The data resulting from analyzing the 492 curriculum documents (241 mathematics curriculum guides and 251 science curriculum guides) was simply a qualitative categorization, providing a dichotomous variable indicating for each topic its presence or absence at a particular grade. Such dichotomous data, when aggregated, were used in TIMSS-95 to represent an important policy variable – the number of topics intended for coverage at each grade level (see Figure 7.2).

In the United States, since there were no national standards at the time of the study, only state standards, the United States was represented by an

The 1995 TIMSS Curriculum Analysis and Beyond 129

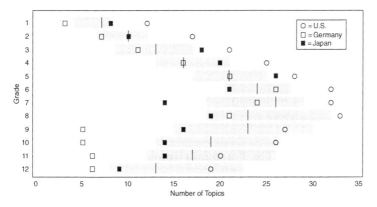

Figure 7.2   Number of mathematics topics intended
Note. Republished with permission of Springer Science and Business Media BV, from "A Splintered Vision: An Investigation of U.S. Science and Mathematics Education," by W. H. Schmidt, C. C. McKnight, & S. A. Raizen, 1997, p. 15; permission conveyed through Copyright Clearance Center, Inc.

aggregate of a PPS random sample of thirty-two states. This aggregate representation of the United States indicated plans to cover so many topics that the US composite shows an intention to cover more mathematics topics than all other countries. This was also characteristic of most states individually. The number of topics to be covered dropped below the 75th percentile internationally only in grades 9, 10, 11, and 12 when the United States typically teaches mathematics in specific courses – algebra, geometry, etc. The gray bars in Figure 7.2 show how many mathematics topics were intended to be covered at each grade in the TIMSS-95 countries. The bars extend from the 25th percentile to the 75th percentile among all participating countries. The black line indicates the median number of topics at each grade. The United States, Germany, and Japan are individually indicated.

This analysis comparing TIMSS-95 countries led to the conclusion that the intended US curriculum was a "mile wide and an inch deep," a policy-relevant observation supported by data showing that the United States expected coverage of many more topics at each grade level than all other countries (Schmidt, McKnight, & Raizen, 1997). This influenced US standards developed after 1997, including the Common Core State Standards (National Governors Association Center for Best Practices & Council of Chief State School Officers, 2010; Schmidt, McKnight, & Raizen, 1997). Other than this use of these data, they mostly served to provide descriptive lists of content coverage, allowing countries to compare their expectations with those of other countries. Given the official nature of these documents and the fact that they represented a policy statement by each country, such resulting comparisons provided benchmarks that countries could use to compare each other. It is in that sense that the TIMSS-95 curriculum analysis became a study providing policy-relevant country comparisons.

Textbooks: The Potentially Implemented Curriculum

It has been fairly well established that textbooks play a major role in classroom instruction and help to define, if not out and out determine, classroom content coverage, especially in mathematics but less so in science (Schiller, Schmidt, Muller, & Houang, 2010; Schmidt, McKnight, Houang, Wang, Wiley, Cogan, & Wolfe, 2001). It has been called the "potentially" implemented curriculum since it is the teacher who finally decides what topics to cover for how long and in what order, but it provides the teacher with one vision of what to cover complete with supporting materials. Textbooks are ubiquitous across the world. In TIMSS-95, we found them to be present in all participating countries but varying in their size and composition (Valverde, Bianchi, Wolfe, Schmidt, & Houang, 2002). The textbook analysis focused on textbooks used in the focal grades (e.g. fourth and eighth) and in the "specialist" courses in the final year of secondary school (e.g. calculus and physics). This led to the analysis of a total of 418 textbooks written in over 20 languages. Country coders were first trained to criterion following a rigid set of rules with performance criteria defining the process.

To analyze this source of data with respect to measuring OTL, a set of document analysis procedures were specified. This included procedures for segmenting the text into what were called "blocks" – sections of the textbook material that dealt with the same topic as defined by the corresponding framework. Each time the topic changed in the textbook flow, a new block was formed. There were two additional dimensions to the framework that provided a richer portrayal of the OTL performance expectations (PE) and perspective (blocks also changed if coding of the PE dimension changed) (Schmidt, McKnight, Valverde, Houang, & Wiley, 1997; Schmidt, Raizen, Britton, Bianchi, & Wolfe, 1997).

Each block was then coded as to the topic using the framework. Multiple topic codes were allowed in the coding but occurred relatively infrequently. The other two dimensions were also coded for each block, producing what was termed the signature of the block, essentially a vector of codes the length of which varied depending on the number of multiple codes for either of the two dimensions.

The resulting coded data were very similar to those derived from the coding of the standards. The difference, however, was that the frequency of such occurrences over blocks for a given topic within a book has meaning, which was not the case with the coded standards. For standards, the frequency of a topic's mention was arbitrary and had no quantitative meaning, only the dichotomous observation of whether or not the topic was present. In textbooks the relative frequency of a topic's mention was taken as an indication of emphasis. This seemed reasonable, given that most textbooks are organized around daily lessons, or at least some segment of classroom time, and classroom time is limited over the year's instruction. This then represents one window on the relative emphasis that potentially could be spent on each topic if the book were to be followed.

The process of coding the textbooks was very labor intensive and time consuming, taking an average of around 120 hours to code a single US mathematics or science textbook. Post TIMSS-95, procedures were developed and used in other non-international studies (Brown, Schiller, Roey, Perkins, Schmidt, & Houang, 2013), which simplified the coding procedures, reducing the time to about forty hours per US textbook. The change in procedure involved the definition of the segment that was to be coded, moving from a "block," which was typically a paragraph of the book, to a segment defined by the book as a lesson; typically around three to five pages in US textbooks. The new procedure was to code the central focus of the lesson. Multiple codings were permitted. Studies of the impact of the change in procedures showed only small differences in the overall estimates of the relative emphasis associated with topic coverage. It had a more significant impact on the PE dimension of the framework.

The textbook-related variables developed in TIMSS-95 to represent OTL were created by first summing the number of "blocks" (or units in later applications) associated with each topic defined by the framework. These were then converted into proportions indicating the relative number of segments associated with each topic. Finally, they were converted into the number of days of instruction by multiplying the proportions by the total number of instructional days in a year for each country.

Textbook data provides one more window on opportunity to learn and how that varies across countries but moves one step closer to classroom instruction and with more information – focusing not just on coverage but also relative emphasis as well as order of presentation. Such data can be used in conjunction with an international assessment to provide a descriptive context to the assessment results as well as to be a quantitative measure of OTL in statistical modeling related to performance.

The TIMSS-95 data clearly showed that to assume what one country covers in mathematics at a given grade is the same as it is in other countries is folly. As a consequence, without such OTL data, interpreting country differences in performance is fraught with the possibility of wrong interpretations. This is always a problem with cross-sectional international data. The measurement of OTL does not automatically make the problem disappear but is helpful if the appropriate OTL is measured relative to the test design.

Table 7.1 illustrates one use of the textbook data, showing the large variation in textbook coverage (proportion of the book) in TIMSS-95 at grade four. The table shows that, for example, in Bulgaria the textbook focused on whole number meaning but no textbook space was allocated to the coverage of fractions, which is almost the complete opposite of textbook coverage in Hong Kong, where the focus is on fractions. Expecting similar performance on a total test score between the two countries would not be reasonable given those differences in OTL.

This is what the founding fathers recognized; performance depends on the content composition of the assessment relative to the content composition of the OTL, in this case as represented by the textbook. Such differences as found in Table 7.1 were characterized and modeled in TIMSS-95 and reported in *Why Schools Matter* (Schmidt, McKnight, Houang, Wang, Wiley, Cogan, & Wolfe, 2001). More subtle analyses were also done, for example characterizing textbooks by the number of topics it took to meet a particular criterion value as to the percent of the textbook (see Tables 7.2 and 7.3 for mathematics and science, respectively). All of these types of analyses can be used to study country differences in the degree of focus, emphasis, and the sequencing of topics.

Topic Trace Mapping

All of the previously described TIMSS-95 OTL measurements related to the intended and potentially implemented curricula (as well as the

Table 7.1 *The share of textbook emphasis for each topic at grade 4*

[Table showing textbook emphasis indicators for mathematics topics across countries. The indices of comparative attention are represented by symbols: □ = 0, small filled square <.35, medium <.65, larger <.95, largest <=1]

Mathematics Topics	Argentina	Australia	Belgium (Fl)	Belgium (Fr)	Bulgaria	Canada	China, People's Republic of	Cyprus	Czech Republic	Denmark	Dominican Republic	France	Germany	Greece	Hong Kong	Hungary	Iceland	Iran	Ireland	Israel
Whole Number: Meaning																				
Whole Number: Operations																				
Common Fractions																				
Decimal Fractions																				
Percentages																				
Estimating Quantity & Size																				
Measurement: Units																				
Measurement: Perimeter, Area & Volume																				
2-D Geometry: Basics																				
2-D Geometry: Polygons & Circles																				
Proportionality: Problems																				
Equations & Inequalities, Graphs & Solutions																				
Data Representation & Analysis																				

Mathematics Topics	Japan	Korea	Latvia	Mexico	Netherlands	New Zealand	Norway	Philippines	Portugal	Romania	Russian Federation	Singapore	Slovak Republic	Slovenia	Spain	Sweden	Switzerland	Tunisia	USA
Whole Number: Meaning																			
Whole Number: Operations																			
Common Fractions																			
Decimal Fractions																			
Percentages																			
Estimating Quantity & Size																			
Measurement: Units																			
Measurement: Perimeter, Area & Volume																			
2-D Geometry: Basics																			
2-D Geometry: Polygons & Circles																			
Proportionality: Problems																			
Equations & Inequalities, Graphs & Solutions																			
Data Representation & Analysis																			

Note: The indices of comparative attention are represented by the following □ = 0 0 <■<.35 .35<=■<.65 .65<=■<.95 .95<=■

Note. Republished with permission of Springer Science and Business Media BV, from "Many Visions, Many Aims: A Cross-National Investigation of Curricular Intentions in School Mathematics," by W. H. Schmidt, C. C. McKnight, G. A. Valverde, R. T. Houang, and D. E. Wiley, 1997, p. 114; permission conveyed through Copyright Clearance Center, Inc.

implemented curriculum described in the next section) characterize only the focal grades. This seems reasonable as the OTL is being characterized for the grade level at which performance is being assessed. Yet the question that arises is whether this is adequate. Some would see such measures collected at the focal grade as adequate for predicting performance on the focal-grade assessment. The problem with that approach, however, is that there are typically no pre-tests in international assessments to at least in part account for prior OTL as reflected in the pretest. As a result, student performance on the assessment is related not only to the content exposure that occurred in the focal grade, but to the cumulative content coverage the student has experienced up to that point in their schooling.

Table 7.2 *Percentage of mathematics textbooks that require a specific number of topics to cover 80 percent of the textbook*

Number of Topics Needed	Population 1 Percentage	Population 2 Percentage	Population 3 Percentage
1		9.7	12.0
2	14.3	8.3	22.0
3	7.1	13.9	16.0
4	21.4	8.3	20.0
5	17.1	8.3	24.0
6	12.9	11.1	2.0
7	10.0	8.3	
8	8.6	2.8	4.0
9	2.9	5.6	
10	4.3	4.2	
11	1.4	2.8	
12		6.9	
13			
14		4.2	
15		2.8	
16		1.4	
17		1.4	
18			
	100.0	100.0	100.0

Note. Republished with permission of Springer Science and Business Media BV, from "According to the book: Using TIMSS to investigate the translation of policy into practice through the world of textbooks," by G. A. Valverde, L. J. Bianchi, R. G. Wolfe, W. H. Schmidt, & R. T. Houang, 2002, p. 97; permission conveyed through Copyright Clearance Center, Inc.

In the original Third International Mathematics and Science Study the sampling design called for the assessment to be given in the two adjacent grade levels defined by Populations 1 and 2 (typically grades 3/4 and 7/8). This was done so that at the country and classroom level the third-grade results could be used as a pre-test for fourth-grade performance (the focal grade of the study), and similarly with grades 7 and 8. The details of this design are discussed in Chapter 9. This unique design justified the use of the focal-grade OTL to be on the same grade as the one in which the focal-grade assessment was given. In prior IEA studies as well as subsequent ones (with the exception of the inclusion of the longitudinal study in SIMS) the use of the focal-grade OTL to predict performance would be subject to

Table 7.3 *Percentage of science textbooks that require a specific number of topics to cover 80 percent of the textbook*

Number of Topics Needed	Population 1 Percentage	Population 2 Percentage	Population 3 Percentage
1	1.7	8.3	2.3
2	6.7	7.5	9.1
3	16.7	13.3	4.5
4	11.7	12.5	11.4
5	10.0	10.0	22.7
6	10.0	15.0	22.7
7	6.7	7.5	9.1
8	8.3	5.8	13.6
9	10.0	5.8	2.3
10	5.0	4.2	
11	5.0	2.5	2.3
12	1.7	0.8	
13	1.7	0.8	
14	1.7		
15	3.3	0.8	
16		0.8	
17		1.7	
18			
19			
20		1.7	
21		0.8	
	100.0	100.0	100.0

Note. Republished with permission of Springer Science and Business Media BV, from "According to the book: Using TIMSS to investigate the translation of policy into practice through the world of textbooks," by G. A. Valverde, L. J. Bianchi, R. G. Wolfe, W. H. Schmidt, & R. T. Houang, 2002, p. 101; permission conveyed through Copyright Clearance Center, Inc.

the criticism that such results would be biased given the absence of a pre-test.

As described in an earlier section of this chapter, the standards documents varied across countries, especially in the level of detail and, as such, only the guides for the upper of the two grades (fourth and eighth) were analyzed. The textbooks were very detailed but, like the OTL collected from teachers, were focused only on the tested grade levels. The task of analyzing textbooks from all previous grades was cost-prohibitive and would be virtually impossible to accomplish.

Given the importance of having such contextual data for all countries at the same level of specificity, SMSO designed an additional procedure, which turned out to be much less prohibitive in cost and time but had great face validity and provided detailed portrayals of each country's K–12 curriculum that turned out to be valuable beyond its original purpose. This procedure was called topic trace mapping.

The TIMSS-95 frameworks were used to develop, for each of mathematics and science, a checklist to be used by each country's curriculum specialist (at the ministry level) to map the coverage of each of the corresponding framework topics over grades K–12. The coding required that for each topic the curriculum specialist used whatever official documents, including textbooks, were necessary to indicate three things:

(1) the grade level at which the topic was first introduced,
(2) the grade level at which coverage of the topic was finished,
(3) the grade or grade levels at which the topic received its greatest focus (more than one focus was allowed).

From these three indices, a coverage map was developed for all countries and for all topics in the framework. Table 7.4 illustrates the resulting topic trace maps for the topics' congruence and similarity, and equations and formulas for a selected set of countries.

The data from the mathematics topic trace mapping proved to be especially valuable as they formed the basis for the development of the International Grade Placement index (IGP). For each topic and across all countries, combining the beginning and ending points with the grade level (s) at which the topic received its focus, an index was derived that statistically placed the topic at a particular grade level from an international point of view.

Given the hierarchical nature of mathematics (such an index was not developed for science) it was assumed that topics covered at the higher grades would be mathematically more complex and as such the index could be viewed as providing a measure of the difficulty of the mathematics coverage found in the standards, textbooks, or covered by teachers at each particular grade level (see Table 7.5 for a listing of the IGP values for a sample set of topics).

For example, countries with a higher value for an aggregate IGP averaged across the individual IGP values for each of the topics covered by teachers, standards, or textbooks at a given grade level or cumulatively up to that grade level were considered to have covered more advanced mathematics and as such might be expected to do better on the assessment.

Table 7.4 *Two representative topic trace maps for a selected set of countries*

	Congruence and Similarity												Equations and Formulas											
	Grades												Grades											
	1	2	3	4	5	6	7	8	9	10	11	12	1	2	3	4	5	6	7	8	9	10	11	12
Argentina	−																				+	+	−	
Canada		−	−	−		−	+	+	+	+	+	−							+	+	+	+	+	
Cyprus					−	−	+	−	−	+	+	−	−			−		−	+	+	+	+	+	
Denmark			−			−	+	+	−	+	+	−				+		−	+	+	+	+	+	
Hungary				−			−	+	−	−	+	−			−		−		+	−	−	+	−	
Iceland							−	+	−	−	+	−								−	+	−	+	
Iran						+		−	+	−	−							+		−	−	−		
Ireland						+	−	+	+	+		−	−		+				+	−	+	+	+	+
Japan					+	+	+	−	+	+	+			+	+		+	+	+	+	+	+	+	
New Zealand							−	−		−	−						−		−	+	+	+	+	
Spain						+	+													−		−	−	
Tunisia							−	+	−	−		−							+	+	−	+	+	+
US						−		+	−	+	+							+	+	+	−	+	+	+

Note. Republished and adapted with permission of Springer Science and Business Media BV, from "Many Visions, Many Aims: A Cross-National Investigation of Curricular Intentions in School Mathematics," by W. H. Schmidt, C. C. McKnight, G. A. Valverde, R. T. Houang, and D. E. Wiley, 1997, p. 194; permission conveyed through Copyright Clearance Center, Inc.

Table 7.5 *International Grade Placement (IGP) values for selected topics*

Selected mathematics topics	IGP
Fractions and Number Sense	
Whole numbers - including place value, factoring, and operations	1.7
Understanding and representing common fractions	4.4
Computations with common fractions	4.4
Simple computations with negative numbers	6.6
Square roots (of perfect squares less than 144), small integer exponents	7.5
Geometry	
Congruence and similarity	8.4
Symmetry and transformations (reflection and rotation)	7.1
Algebra	
Simple algebraic expressions	7
Representing situations algebraically; formulas	7

Note. From "Equality of Educational Opportunity: Myth or Reality in U.S. Schooling?," by W. H. Schmidt, L. S. Cogan, and C. C. McKnight, American Educator, Winter 2010–2011, p. 14. Copyright by American Educator.[1]

This index proved to have strong face and construct validity and was used in numerous studies relating IGP, as a measure of OTL, to performance (see Schmidt et al., 2001; Schmidt, Cogan, & Guo, in press; Schmidt, Cogan, Houang, & McKnight, 2011; Schmidt & Houang, 2007). The IGP in these applications can be thought of as a set of weights that can be multiplied by the dichotomous variable characterizing coverage (0, 1) for each topic or by the amount of time a teacher or textbook spends on the coverage of each topic in the framework, resulting in an aggregate estimate of coverage or the relative amount of such coverage.

Perhaps one of the most striking and most referred to results of topic tracing other than the IGP was its use in the development of the A+ profile, an international benchmark defining the K–8 curriculum of the top-achieving countries in TIMSS-95. The development of the A+ benchmark was created using the data that resulted from the topic trace mapping for the top achieving countries. In mathematics, five countries were chosen: Japan, Korea, Belgium (Flemish part), Singapore, and Hong Kong. These were the top-performing countries, whose scores

---

[1] Portions reprinted with permission from the Winter 2010–2011 issue of *American Educator*, the quarterly journal of the American Federation of Teachers, AFL-CIO.

were substantially above those of other countries on the eighth-grade assessment. Topics were included in the benchmark at a grade where two-thirds or more of the five countries covered that topic at that grade level. In this way, the A+ benchmark represented a composite of these five countries and did not represent any one of them individually.

For science the same procedure was followed but there were only four countries that met the criterion of top-achieving. This included Hong Kong, Japan, Korea, and Singapore. Because of the number of countries, the cutoff for the placement of a topic at a specific grade level was even more strict – 75 percent. These benchmarks, represented in Figures 7.3 and 7.4 for both mathematics and science, have been extensively used worldwide in the revision of various national standards, including the United States and its Common Core State Standards for Mathematics (CCSSM).

More directly, in its use in the assessment realm, the A+ has been found to be statistically significantly related to country performance based on the similarity of a country's own K–8 curriculum map to that of the international benchmark (see Schmidt & Houang, 2007). Separately, it has also been used to examine the relationship of the new Common Core State Standards to the A+ benchmark in the United States (Schmidt & Houang, 2012).

Growing out of the benchmark analysis, two concepts were identified that characterized the top-achieving countries. Focus referred to the fact that top-achieving countries on average intended to cover fewer topics per grade level than did other countries. The resulting index was either the total number of topics intended over grades K–8 or the total by each grade level separately. Figure 7.2 speaks to the issue of focus.

The second characteristic that emerged was coherence. This recognized that topics are not arbitrary as to their placement over the grades in the curriculum of a country. Because mathematics is hierarchical in nature, this is especially true, but to define the index, the sequencing structure implied by the A+ benchmark was used to define coherence and hence define the desirable sequencing of the topics over grades. The fact that the A+ for both mathematics and science represented in Figures 7.3 and 7.4 are like upper triangles provides a visual that reflects the principle of coherence. Three OTL indices characterizing coherence were developed for analytical purposes. All three were based on the A+ benchmark, using it as a template:

Topic	1	2	3	4	5	6	7	8
Whole Number Meaning	●	●	●	●	●			
Whole Number Operations	●	●	●	●	●			
Measurement Units	●	●	●	●	●	●	●	
Fractions			●	●	●	●		
Equations & Formulas				●	●	●	●	●
Data Representation & Analysis			●	●	●	●		●
2-D Geometry Basics			●	●	●	●	●	●
Polygons & Circles				●	●	●	●	●
Perimeter, Area & Volume				●	●	●	●	●
Rounding & Significant Figures					●	●		
Estimating Computations					●	●	●	
Properties of Whole Numbers Operations					●	●		
Estimating Quantity & Size					●	●		
Decimals					●	●	●	
Relation of Decimals & Fractions					●	●	●	
Properties of Decimals & Fractions						●	●	
Percentages						●	●	
Proportionality Concepts						●	●	●
Proportionality Problems						●	●	●
2-D Coordinate Geometry					●	●	●	●
Geometric Transformations							●	●
Negative Numbers, Integers & Their Properties							●	
Number Theory							●	●
Exponents, Roots & Radicals							●	●
Orders of Magnitude							●	●
Measurement Estimation & Errors							●	
Constructions Using Straightedge & Compass							●	●
3-D Geometry							●	●
Congruence & Similarity								●
Rational Numbers & Their Properties								●
Functions								●
Slope								●

Intended by two-thirds or more of the top-achieving countries ●

Figure 7.3 Mathematics topics intended at each grade by top-achieving countries

(1) Topics covered early was represented by a count of the number of topics covered prior to when they were covered in the A+.
(2) Topics covered later was similarly defined for topics covered at later grades than specified in the A+.
(3) Topics covered at the same grades as the A+ were tabulated by a simple count.

The 1995 TIMSS Curriculum Analysis and Beyond

Topic	1	2	3	4	5	6	7	8
Organs, Tissues	●	●	●	●	●	●		
Physical Properties of Matter	●	●	●	●	●	●		
Plants, Fungi	●	●	●	●	●	⊙		
Animals	●	●	●	●	⊙	●		
Classification of Matter	⊙	⊙	⊙	⊙	●	●		
Rocks, Soil	⊙	⊙	⊙	⊙	●	●		
Light	⊙				●	●		
Electricity		⊙		⊙	●	●		
Life Cycles		●	●	●	●	●		
Physical Changes of Matter		●	●	●	●	●		
Heat & Temperature		●	●	●	●	●		
Bodies of Water		⊙	⊙	⊙	●	●		
Interdependence of Life				⊙	●	⊙	⊙	
Habitats & Niches				⊙	⊙	⊙	⊙	
Biomes & Ecosystems				⊙	●	⊙	⊙	
Reproduction				⊙			⊙	
Time, Space, Motion				●	●	●	●	
Types of Forces				⊙	⊙	●	●	
Weather & Climate				⊙	⊙	●	●	
Planets in the Solar System				⊙	⊙	⊙	⊙	
Magnetism					●	●	●	
Earth's Composition					⊙	●	●	
Organism Energy Handling					⊙	⊙	●	
Land, Water, Sea Resource Conservation					⊙	⊙	●	
Earth in the Solar System					⊙	⊙	⊙	
Atoms, Ions, Molecules							●	●
Chemical Properties of Matter							●	●
Chemical Changes of Matter							●	●
Physical Cycles							⊙	●
Land Forms							⊙	●
Material & Energy Resource Conservation							⊙	●
Explanations of Physical Changes							⊙	⊙
Pollution							⊙	●
Atmosphere							⊙	⊙
Sound & Vibration							⊙	⊙
Cells							⊙	⊙
Human Nutrition							⊙	⊙
Building & Breaking								●
Energy Types, Sources, Conversions								●
Dynamics of Motion								⊙
Organism Sensing & Responding								⊙
Number of additional topics intended, on average, by A+ countries to complete their curriculum at each grade level.				5/11	5/13	3/15	2/19	4	4

Intended by all but one of the A+ countries (3 out of 4). ⊙
Intended by all of the A+ countries. ●

Figure 7.4 Science topics intended at each grade by a majority of top-achieving countries
Note. Adapted from "Curriculum coherence: An examination of US mathematics and science content standards from an international perspective," by W. H. Schmidt, H.- C.- Wang, & C. C. McKnight, 2005, p. 545. Permission granted by Copyright Clearance Center Inc (CCC) on Taylor and Francis's behalf.

These three indices could be calculated for all countries and then used in various analyses. Table 7.6 shows the results of such a set of analyses for twenty to thirty-three countries, relating indices of coherence and focus to performance. The analyses show these indices are statistically significantly related to TIMSS-95 performance at grades 7 and 8, but only marginally for grade 4. The advantage of these indices and relating them to performance is that they are cumulative and as a result can be used with cross-sectional data.

The invention of the topic trace mapping procedure though simple in conception and easy to interpret, produced a powerful quantitative methodology, that was innovative and strikingly revealing for the measurement of OTL. In analysis after analysis the use of IGP in conjunction with OTL has proved to be a powerful predictor of performance (Cogan, Schmidt, & Wiley, 2001; Schmidt, Cogan, Houang, & McKnight, 2011; Schmidt et al., 2001). Being quantitative in nature it has advanced the measurement of OTL and, in that sense, given the field a measure of schooling that has the validity and reliability to stand up in analyses pitting home against schooling in explaining performance. The A+ benchmark also derived from topic trace mapping data has proven to be exceptionally valuable toward having an empirically derived international benchmark around curriculum.

Implemented Curriculum

In the final analysis, it is what content is actually covered in the classroom, for how much time, in what order, and the time students are actively engaged in instruction that has the greatest effect in the aggregate on what students learn. This is true logically, even though empirically it is difficult to establish, especially in cross-sectional international assessments. Schmidt, Jorde, et al. (1996) stated the importance of classroom instruction in this way, "Classroom activities are a dynamic interaction between subject matter content, teachers and students. Teachers use broader curriculum goals and intentions to select or create pedagogical activities and exercises packaged into lessons for students" (p. 70).

Opportunity to learn, as defined by government institutions through standards, represents the official intentions of the country and as such is an important policy context in which to interpret country-level performance. If geometric projections, for example, is not included in the standards for

Table 7.6 Regression analyses relating coherence and focus to achievement

Predictor	Grade 3			Grade 4			Grade 7			Grade 8		
	Estimate	Std error	p	Estimate	Std error	p	Estimate	Std error	p	Estimate	Std error	p
Model												
Number of Topic/Grade Combinations Aligned with Ideal Scenario	9.94	7.38	0.196	7.48	4.04	0.081	3.57	1.48	0.023	2.83	1.12	0.017
Total Number of Topic/Grade Combinations in Curriculum	−2.42	1.13	0.047	−2.58	1.05	0.025	−1.64	0.57	0.008	−1.39	0.45	0.004
Model Fit												
R-Square	0.2191			0.2574			0.2205			0.249		
Residual Mean Squares	1652.9			1489.0			1476.0			1331.4		
p<	0.1222			0.0687			0.0270			0.0136		
Number of Countries	20			21			32			33		
Standardized Coefficient												
Number of Topic/Grade Combinations Aligned with Ideal Scenario	0.4374			0.7331			0.8481			0.8240		
Total Number of Topic/Grade Combinations in Curriculum	−0.6969			−0.9703			−1.0044			−1.0204		

Note. From "Lack of Focus in Mathematics Curriculum," by William H. Schmidt & Richard T. Houang. In T. Loveless (Ed.), 2007, *Lessons learned: What international assessments tell us about math achievement*, p. 76. Copyright 2007 by Brookings Institution Press.

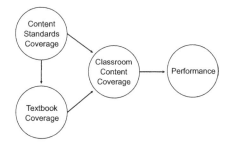

Figure 7.5   A hypothesized structure for curriculum

grade eight for a country then the country performance results can be used to address the wisdom of that curriculum policy.

On the other hand, what content is represented in the country's textbooks comes closer to what is actually taught and in that sense elucidates the country's intentions. Standards are abstractions representing the country's curriculum policies; the textbooks translate those into what activities can be done in the classroom. In that sense, they define the opportunities that might be, and therefore can have either a direct or an indirect effect on performance. The indirect effect comes through classroom coverage. This is reflected in Figure 7.5, but, as it indicates, it is at the classroom level – either at a specific grade or cumulatively over multiples grades – that OTL has its direct effect on performance. The question remains, does this represent the reality of instruction in schools? We look at this later in the chapter.

As suggested previously, there are three indicators of OTL that can be measured: topic coverage (yes/no), amount of instructional time devoted to a particular topic, and the sequence in which the topics are presented. The first two were measured in TIMSS-95, while the sequencing of topics, although it could be done, has not been incorporated in TIMSS-95 analyses.

An entirely different methodology than those described previously in this chapter was needed to characterize OTL at the classroom level. There are basically two informants to specify what content is actually covered – teachers and students. Both TIMSS-95 and PISA, as described in Chapter 3, employ a two-stage PPS random sampling design, with schools being the unit in the first stage. What differs and has implications for measuring the implemented curriculum is the sampling unit in the second stage. TIMSS-95 chose the teacher as the informant, since the classroom was the primary sampling unit in the second stage. PISA, on the other hand, chose the student for the parallel reason that the student was the sampling unit in the second stage.

Using the two frameworks at the higher levels of abstraction, around twenty topics were chosen for each of the science and mathematics TIMSS-95 teacher questionnaires. The topics were chosen to be mutually exclusive and exhaustive of the K–8 curriculum. To allow for cross-national differences in what topics were covered at each grade, topic selection was centered at the focal grade plus or minus one using the TIMSS-95 curriculum data (from the standards, textbooks, and topic trace mapping analyses) that had been collected and analyzed two years prior to the main TIMSS-95 data collection. It is important to note that in TIMSS-95 the focus of the OTL measurement is not at the item level but at the topic level, defining categories of content as derived from the two frameworks.

Some of the topics chosen to be in the teacher questionnaires were from higher aggregate levels and were therefore inclusive of several subtopics. In other cases, for content that was focal for a particular grade, more detail and specificity from the lower levels of the framework were used. Additionally, in developing the teacher questionnaires, the language was changed from the mathematics-oriented definitions used in the framework to language more "user friendly" for teachers. The same was done for the science teacher questionnaire.

The question asked on the teacher questionnaire for each of the included topics was, "choose the category (from any 5) that indicates the number of periods you will spend on this topic over the school year" (see Appendix B for a copy of the questionnaire). The teachers were given the questionnaire in the spring when the assessment was being administered. The data from the OTL questionnaires were first converted to a proportional representation of the number of periods allocated by a teacher to each of around twenty topics. This was then converted to an estimate of the amount of time over the year given to coverage of each topic. The conversion for US data was based on the assumption of 160 actual instructional days in a year.

Using this estimate of the amount of time allocated to each of the grade-relevant topics was found to be a reliable and valid indicator of OTL and of its relationship to performance on the Third International Mathematics and Science Study assessment at all levels of analysis – country, classroom, and student (Schmidt et al., 2001). This was possible given the cohort-longitudinal design used in TIMSS-95. For example, statistical models relating teacher OTL to performance were statistically significant with large coefficients (Schmidt et al., 2001). Examples of these results will be discussed in more detail in Chapter 10.

The Structure of Curriculum

The structure of curriculum is really the question of how do the intended, potentially implemented, and implemented versions of OTL fit together to define the curriculum as experienced by a classroom of students, and how is that curriculum as presented related to what students learned? We now return to Figure 7.5 to suggest a hypothetical structure of curriculum and how it is related to learning (gain). Gain is defined in the original TIMSS-95 study analyses as the pseudo-gain of the average eighth-grade score minus the average seventh-grade score for a country. Figure 7.6 shows three different country structures as estimated from the data.

Japan's estimated model suggests a pattern where the intended, potentially implemented, and implemented all line up with the hypothetical model found in Figure 7.5. In the fitted model the standards defining the intended curriculum not only influenced the textbooks and classroom coverage directly but also had an indirect relationship to classroom coverage through its relationship to textbook coverage. The implemented curriculum characterized by instructional time was also statistically significantly related to the gain score. Japan is an example of a well-aligned system in which all parts work coherently toward student learning. The other two countries in Figure 7.6 exhibit different structures. For Slovenia, the pattern is much the same as Japan but with one notable exception: the absence of the country's content standards bearing any relationship to the other parts of the system. For Slovenia, the textbooks serve as the sole driving force for the system.

Finally, the US results show a more chaotic pattern or structure. The textbooks drove the outcome, as was the case in Slovenia. The absence of a relationship of the standards to any other part of the system is understandable, as at that time there were no widely accepted or national standards like the Common Core State Standards. The analysis used an aggregate of thirty-two states' standards. The nature of the relationships of textbooks to other parts of the system is what is more chaotic, as it is related to instructional time, but unlike the other two countries, instruction has no direct relationship to gain, which one would expect. Textbooks did, however, have a direct relationship to gain, which one would not expect given the hypothesized model in

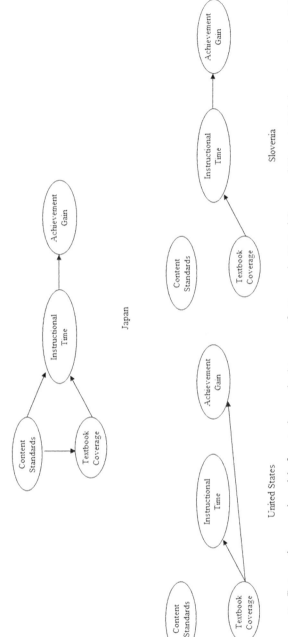

Figure 7.6 Estimated structural model of curriculum and achievement for Japan, the United States, and Slovenia (eighth-grade mathematics)
Note. Arrows indicate statistically significant positive relationships; absence indicates no relationship. Republished with permission of John Wiley and Sons Inc., from "Why Schools Matter: A Cross-National Comparison of Curriculum and Learning," by W. H. Schmidt, C. C. McKnight, R. T. Houang, H. Wang, D. E. Wiley, L. S. Cogan, & R. G. Wolfe, 2001, pp. 283–284; permission conveyed through Copyright Clearance Center, Inc.

Figure 7.5. The conclusion one is drawn to is consistent with what has been said about US education in mathematics – it is driven by the textbook – and that appears to not go through the teacher and classroom instruction.

We believe these results, once again, reinforce the foresight the founding fathers had as to why the use of international assessments in education had to be accompanied by OTL measures. The original TIMSS-95 study provided the machinery to realize their goal and to show how important the resulting data are to the proper interpretation of the assessment results. In that way, the Third International Mathematics and Science Study broke new ground, providing new insight into how curricular policy and its consequences are not only important to study in their own right but also in their relationship to academic performance.

Trends in International Mathematics and Science Studies

Over the twenty-plus years following TIMSS-95, IEA has had a mathematics and science study every four years – 1999, 2003, 2007, 2011, and, most recently as of the writing of this book, 2015. In each of those five studies, the IEA OTL tradition was continued but in a greatly reduced manner from that in SIMS and TIMSS-95. Table 7.7 provides a few examples of the teacher and curriculum expert OTL measures that have been included in these studies. Here we review how OTL was measured in the most recent 2015 study.

Intended Curriculum

The Test-Curriculum Match Analysis has been conducted with mathematics and science experts prior to each TRENDS administration since 1995. As described in Chapter 5, experts from each country responded to each test item, noting whether the content represented in the item was in the country's curriculum for students at the grade to be tested. This has been viewed primarily from the perspective of ensuring the validity of the assessment for students in each country. Nonetheless, as experts are expected to make use of their curriculum standards/documents in making this determination, this exercise also provides a window on the intended curriculum but only for a specific grade.

Beginning with the 2003 TRENDS assessment cycle, a Curriculum Questionnaire was introduced. There were four versions: one for each academic area, e.g. mathematics and science, and grade to be assessed,

Table 7.7 Examples of OTL measurement in IEA studies from 1999 TIMSS-R and TRENDS

Focus	Prompt	Question	Reference

Exposure/Coverage Teachers	Topics	1. The following list includes the main topics addressed by the TIMSS mathematics test. Check the response that describes when students in your mathematics class have been taught each topic. If a topic has been taught before this year and also in the current year, check the two boxes that apply. Otherwise, check one box in each row.
		a. Taught before this year
		b. Taught 1–5 periods this year
		c. Taught more than 5 periods this year
		d. Not yet taught
		e. I do not know
		Examples of topics listed:
		a. Fractions and Number Sense
		  1. Whole numbers – including place values, factorization and operations (+,-,x,÷)
		  2. Understanding and representing common fractions
			TIMSS-R 1999 Mathematics Teacher Questionnaire TIMSS web site

Emphasis in Instruction Teachers	Topics	What subject matter do you emphasize <u>most</u> in your mathematics class? Check only one: a. Mainly number (e.g. whole numbers, fractions, decimals, percentages, etc.). b. geometry. c. algebra. d. combined algebra and geometry. e. combined algebra, geometry, number, etc. f. other, please specify _____	**TIMSS-R 1999** Mathematics Teacher Questionnaire TIMSS web site
Exposure/Coverage	Topics	1. The following list includes the main topics addressed by the TIMSS mathematics test. Choose the response that describes when students in	**TRENDS 2003**

Table 7.7 (cont.)

Focus	Prompt	Question	Reference

8th Grade Teachers		your mathematics class have been taught each topic. If a topic was taught half this year and half before this year, please choose "Mostly taught this year."
		a. Mostly taught before this year
		b. Mostly taught this year
		c. Not yet taught or just introduced
		Examples of topics listed:
		A. Number
		  a. Whole numbers including place values, factorization and the four operations
		  b. Computations, estimations, or approximations involving whole numbers
			Mathematics Teacher Questionnaire TIMSS web site

Emphasis in Instruction 8th Grade Teachers	Topics	By the end of this school year, approximately what percentage of teaching time will you have spent during this school year on each of the following mathematics content areas for the TIMSS class? Write in the percent. The total should add up to 100%: a. Number (e.g. whole numbers, fractions, decimals, ratio, proportion, percent). b. Geometry (e.g. lines and angles, shapes, congruence and similarity, spatial relationships, symmetry and transformations). c. Algebra (e.g. patterns, equations and formulas, relationships). d. Data (e.g. data collection and organization, data representation, data interpretation, probability). e. Measurement (e.g. attributes and units, tools, techniques and formulas). f. other, please specify_____	TRENDS-2003 Mathematics Teacher Questionnaire TIMSS web site

Table 7.7 (cont.)

Focus	Prompt	Question	Reference
Exposure/Coverage Curriculum Experts	Topics	According to the national mathematics curriculum, what proportion of grade 8 students should have been taught each of the following topics or skills by the end of grade 8? Across grades K–12, at what grade(s) are the topics primarily intended to be taught? Proportion of grade 8 students expected to be taught topic: Not included in the curriculum through grade 8 Only the more able students (top track) All or almost all students Grade(s) topic is expected to be taught K–12: (*grade(s) to be written in*) Examples of topics listed: A. Number a. Whole numbers including place values, factorization and the four operations b. Computations, estimations, or approximations involving whole numbers	**TRENDS 2003** Curriculum Questionnaire TIMSS web site
Exposure/Coverage Curriculum Experts	Topics	According to the national mathematics curriculum, what proportion of grade 8 students should have been taught each of the following topics or skills by the end of grade 8? Across grades from preprimary through upper secondary education, at what grade(s) are the topics primarily intended to be taught? Proportion of grade 8 students expected to be taught topic: All or almost all students Only the more able students Not included in the curriculum through grade 8 Grade(s) topic is expected to be taught K–12: PP G1 G2 G3 G4 G5 G6 G7 G8 G9 G10 G11 G12	**TRENDS 2015** Curriculum Questionnaire TIMSS web site

Table 7.7 (cont.)

Focus	Prompt	Question	Reference

		Examples of topics listed:
		A. Number
		  a. Whole numbers including place values, factorization and the four operations
		  b. Computations, estimations, or approximations involving whole numbers	

Exposure/Coverage
8th Grade Teachers
	Topics	The following list includes the main topics addressed by the TIMSS mathematics test. Choose the response that best describes when students in this class have been taught each topic. If a topic was in the curriculum before the eighth grade, please choose "Mostly taught before this year." If a topic was taught half this year but not yet completed, please choose "Mostly taught this year." If a topic is not in the curriculum, please choose "Not yet taught or just introduced."
		  a. Mostly taught before this year
		  b. Mostly taught this year
		  c. Not yet taught or just introduced
		Examples of topics listed:
		A. Number
		  a. Computing with whole numbers
		  b. Comparing and ordering rational numbers
		  . . .
		  e. problem solving involving percents or proportions
			TRENDS-2015 Mathematics Teacher Questionnaire TIMSS web site

e.g. typically fourth and eighth. These questionnaires have multiple parts addressing issues around the national curriculum, e.g. is there one; who is responsible for it; pedagogical approaches; calculators and computers usage; teacher education and certification; and eighth-grade topics in the national curriculum (or fourth grade in the fourth-grade version of the survey). Other ancillary issues have been included in subsequent cycles.

The portion of the Curriculum Questionnaire that most directly addressed the *intended* aspect of the IEA curriculum model includes two questions posed about the national mathematics curriculum. The first asks "... what proportion of grade 8 students should have been taught each of the following topics or skills by the end of grade 8?" The second asks, "[a]cross grades from preprimary through upper secondary education, at what grade(s) are the topics primarily intended to be taught?" (see Table 7.7). The topics listed in the Curriculum Questionnaire are the same eighth-grade (or fourth-grade) topics listed in the Teacher Questionnaire (see Appendix D). This does provide a cross-grade window on those grade-specific topics, yet only those eighth (or fourth) grade topics on the test. Consequently, it fails to provide the broader context necessary to understand the complete curriculum at either of these two grades as it does not include those topics covered by a country's curriculum that are not included on the TRENDS assessments.

Furthermore, it does not provide a portrait of the country's complete K–8 curriculum as it only addresses the tested topics. This approach provides additional information related to the Test-Curriculum Match Analysis (TCMA). What it does not provide is a broader context of a country's whole K–8 curriculum and, given the hierarchical nature of mathematics, where one set of topics provides the foundation for another set of more advanced topics, understanding those relationships is critical to providing a more complete understanding of the relationship of OTL to achievement, especially given the absence of a pre-test. The full list of twenty eighth-grade mathematics topics from the 2015 Mathematics Curriculum Questionnaire is provided in Appendix D.

As noted earlier, this issue of context – what else is in the curriculum for the tested grade as well as in the prior grades and the degree of emphasis they receive – is critical. For example, one eighth-grade number topic is "computations, estimations, or approximations involving whole numbers" (see Table 7.7). This may be a major focus in eighth grade in one country, much like it is a major focus in grade four in another, or it may be addressed in an ancillary manner as part of a rigorous algebra-focused curriculum. The footnote to the TCMA noted earlier in Chapter 5 is entirely relevant here in terms of the caution noted: Because a country's curriculum may include topics other than those listed, there is no way to determine the extent to which the Curriculum Questionnaire reflects the full curriculum of the country (see page F-1, Mullis et al., 2016). Consequently, the full context, focus, and emphasis that were captured in the TIMSS-95 Topic Trace Mapping analysis cannot be determined (Schmidt, Houang, & Cogan, 2002; Schmidt, McKnight, Valverde, Houang, & Wiley, 1997; Schmidt, Raizen, Britton, Bianchi, & Wolfe, 1997; Schmidt, Wang, & McKnight, 2005; Valverde & Schmidt, 2000).

Implemented Curriculum

TRENDS has also continued to include a measure of the implemented curriculum in each subject- and grade-specific version of the teacher questionnaire. Teachers are asked to indicate one of three possible responses to a list of topics. For example, the question on the eighth-grade mathematics teacher questionnaire indicates that the "list includes the main topics addressed by the TIMSS mathematics test." (see Table 7.7). Teachers are to indicate whether the topic was "mostly taught before this year"; "mostly taught this year"; or "not yet taught or just introduced" (see Table 7.7 and Appendix D). A substantial deviation from the Third International Mathematics and Science Study is the level of detail with respect to the option, "taught this year." TIMSS-95 asked teachers to indicate how frequently each topic was taught during the year of the study in terms of an estimate of the number of periods spent covering that topic, whereas the TRENDS surveys merely request an indication of the year the topic had been taught, i.e. before the current year, during the current year, or later. The effect of this difference is that the TRENDS studies yield only a dichotomous indication of whether the topic was covered during the year of the study that lacks any sense of instructional emphasis.

In addition, the teacher topics in TIMSS-95 were a mutually exclusive and exhaustive list of topics that may be taught. TRENDS-2015, however, constrained the teacher topics to those represented on the student assessment. This constraint on the *implemented* curriculum is the same as that noted for the *intended* curriculum discussed with respect to the Curriculum Questionnaire, with similar consequences.

As has been noted in this chapter and others, without reasonable measures of OTL, international assessments essentially become a cognitive

Olympics – the goal of which is to crown the one who ends up in first place rather than to learn anything about the role of schooling. The former is the work of showmen; the latter is the work of scholars. Both can be appreciated but for different purposes. Both PISA and the IEA studies strive to address the latter purpose, although not always as successfully as many would wish.

CHAPTER 8

Characterizing Student Home and Family Background

As discussed in Chapter 4, international comparative studies have evolved to regularly produce "cognitive Olympics" tables of ranked average country performance (Husén, 1974). As related in Chapters 1 and 2 in reviewing the historical development of these studies, the motivation researchers had in initiating international comparative studies was to exploit the phenomena of naturally occurring educational experiments (different countries have different education systems) to learn more about students' school-related learning (Torney-Purta & Schwille, 1986).

Given this interest in and focus on the effect of schooling on student achievement, the fact that socioeconomic status (SES) as a concept is ubiquitous across all international educational assessments and measures of it are considered absolutely essential might seem odd. On the other hand, Carnoy, Khaveson, and Ivanova (2015) have recently noted, "Many studies have shown that various proxy measures of students' family academic resources, such as mother's education, parents' education, and books or articles in the home, are correlated with students' academic achievement" (p. 250).

A review of how family background factors have been measured in international comparative education studies identifies three important rationales for including such factors (Buchmann, 2002). One reason to include family background measures is simply the need to statistically control for social class in order to be able to obtain an unbiased estimate of the effect of schooling on student learning. There is no conceptual interest in the effect of SES; its only purpose is statistical control.

This point needs to be further developed as it is central to the theory not only of how SES is related to achievement as noted by Buchmann but also as to how it is related to measures of schooling such as OTL and the interplay between all three. This is critical to the study of schooling as found in TIMSS (both TIMSS-95 and TRENDS) and PISA studies. With the exception of the SIMS longitudinal study and TIMSS-95 with its cohort longitudinal design, other IEA and PISA studies are cross-sectional in design and without a pre-test. In such studies the achievement measured at the end of the year is partly, if not mostly, a reflection of the achievement that would have been measured at the beginning of the year if there were a pre-test. The achievement at the beginning of the year would be highly correlated with SES, so in the prediction of end-of-year achievement, SES becomes the surrogate for the pretest.

This is especially true if the focus of the study is on schooling by trying to estimate the effect of within-year factors, such as OTL. If we had had a pre-test (as in SIMS), the SES effects would be much reduced and might refer directly just to how home factors affect within-year learning and indirectly through mediation to how instruction and OTL are differentially allocated. Also, because SES is an imperfect indicator of the pre-test, the effects of those within-year variables are mis-specified and biased.

A second and related reason for the inclusion of SES in international studies, according to Buchmann (2002), is that there is a conceptual interest in the nature of how SES is related to student exposure to content, not just what is learned from the experience. The child development research literature is replete with studies that have focused attention on discovering the effects of a wide range of family and home factors on student learning opportunities and experiences. This rationale was summarized by noting that "[b]irth-weight, school achievement, pubertal timing, body image, family structure, neighborhood cohesion, general quality of life and many other indicators show systematic variation with ethnicity and socioeconomic status" – all factors that are inextricably linked to the family (Entwisle & Astone, 1994, p. 1522). Parents with varied SES backgrounds may have different expectations and perspectives on education that can affect student motivation to learn, which may then in turn affect student responses to in-school learning experiences.

Returning to the TIMSS and PISA studies with their focus on the role of schooling across the globe, this second rationale as noted by Buchmann (2002), is critical to understanding the interplay of OTL, SES, and achievement as it unfolds in the process of schooling. Irrespective of the effect on end-of-year achievement, it is important to understand the relation of SES to resource allocation and the distribution of OTL to various groups of students. Do students with low SES get a fair share of the resources? Do they get equal OTL? Of course, this analysis is also mis-specified and biased because SES is correlated with initial achievement, so there may be good reasons to assign resources and OTL differentially. But in theory the allocation would be according to need, with equal opportunities to learn accompanied with more resources for initially low performers.

As discussed in the last section of this chapter, this does not seem to happen very often.

This implies that we also need to study OTL and other resources as dependent variables, predicted by SES and other background variables. Finally, a simple correlation of end-of-year achievement and SES, expressed perhaps as achievement gaps, without further explanation (forgetting for the moment OTL, etc.) tells us important information about the degree to which equity and social justice is finally attained by the education system.

Finally, a third rationale, according to Buchmann (2002), emerges with international comparative studies of education. So much about education is embedded in cultural and social traditions that vary from one country to another. Many of the personal and family background descriptors such as ethnicity, race, gender, and SES that are frequently included in these studies are social constructs, the meaning and impact, therefore, may not be assumed to be invariant across countries (Entwisle & Astone, 1994). What this means is that the social cultural context is important to understanding cross-country differences. The reasons for including SES here involve understanding the nature of the relationship.

Exploring the specific details of how different family background variables may have different relationships with student performance in different countries is appropriately a focus of anthropological and sociological research and is well beyond the scope of most international comparative education studies. Nonetheless, it is necessary to have at least a rudimentary understanding of key family background variables and how these relate to measures of student SES. In order to fulfill this endeavor, this chapter includes four sections. First we provide an overview of SES and how this has been operationalized. Next we illustrate how SES has been incorporated into education studies and then provide some consideration of SES issues specific to cross-country comparisons. In the third section, we specifically look at how some of the IEA and OECD studies have incorporated various aspects of SES, and then we finally consider some implications for the relationship demonstrated between SES and students' opportunity to learn (OTL).

What Is Socioeconomic Status (SES)?

Imagine that families within a country are stratified or ordered along dimensions of wealth, income, and social status. One of the significant challenges in international work is that where one family is located within their own country's distribution of income, wealth, or status may be quite different given the nature of another country's distribution. In other words, unlike physical measurements such as weight or height, the instruments measuring SES do not, indeed cannot, utilize commonly agreed units of measurement such as meters, feet, inches, pounds, or kilograms. Advances in psychometric methods are able to combine any set of items into a scale that has a defined common unit across any administrative context yet this does not automatically imply that the meaning of a unit difference is invariant across countries; the meaning of the difference in scale units is not always the same in all countries.

The National Center for Educational Statistics (NCES) (2012) report defines SES broadly as "one's access to financial, social, cultural, and human capital resources. Traditionally a student's SES has included, as components, parental educational attainment, parental occupational status, and household or family income" (p. 4). The report also noted that providing such a definition was an important step; yet the definition alone does not imply any particular measurement method or operationalization. Indeed, as the report noted, various studies have adopted different operational definitions. A 1928 report outlined an approach for obtaining a quantitative scale of family SES that consisted of four components: (1) the number of cultural possessions in the home, (2) the home's effective income, (3) family members' participation in community groups, and (4) number of household material possessions (Chapin, 1928). Of particular interest in this SES definition underscoring the contextual and relative nature of the metric is the explicit statement that the status measured is relative to the "average standards ... of the community" (Chapin, 1928, p. 99).

How Is SES Measured?

Measures for each of the "Big 3" SES components – parent's education background, parent's occupation, and family/household income or wealth – originated from the psychological and sociological literature (Buchmann, 2002; NCES, 2012). Each of these components is first measured in order to construct a composite SES index. A number of different international standard scales have been developed for education and occupation from which researchers may choose. Measurement approaches for family income, however, are more diverse.

Occupation

One of the debates in the sociological literature around the measurement of occupation is whether such measurement is to be considered continuous or categorical. The International Standard Classification of Occupations scale (ISCO) developed by the International Labor Organization provides a four-digit code of occupations found in many countries and are often considered as a continuous scale. The International Socioeconomic Index (ISEI) has been developed from the ISCO codings into a categorical scale "to maximize the role of occupation as an intervening variable between education and income" (Ganzeboom, De Graaf, &Treiman, 1992, p. 2). Categorical approaches reflect the view that society is divided into sociologically important discrete classes, while the continuous perspective underscores differences among occupations in greater detail and specificity (Ganzeboom, De Graaf, & Treiman, 1992).

Both the ISCO and the ISEI have demonstrated reliability internationally and have been used in international studies (Buchmann, 2002). The PISA databases have included both the four-digit ISCO occupation codes along with a mapping of these into ISEI categories for both mother's and father's occupation. The PISA index of economic, social, and cultural status (ESCS) has made use of the ISEI categories. Beginning with PISA 2012, the ISCO changed from the 1988 version to the one made available in 2008 (OECD, 2014b). The very early IEA studies did gather some information about parents' occupation but this was not included in TIMSS-95 or in any subsequent studies. The challenge of reliably obtaining this sort of information from student populations, most frequently the focus of the IEA studies, has often been noted (NCES, 2012).

Educational Background

The International Standard Classification of Education (ISCED) was developed by UNESCO in the 1970s to provide categorical descriptions of levels of education that are internationally comparable. It was updated in 1997 and again in 2012 (UNESCO, 1997, 2012). The ISCED levels have been used in IEA and PISA studies. The Comparative Analysis of Social Mobility in Industrial Nations (CASMIN) scale developed at the University of Mannheim in Germany is similar to the ISCED but provides greater distinctions between education and vocational-technical education (Buchmann, 2002). Both of these scales defined levels of education focused on the number of years of schooling required to attain each level.

Wealth

Measures of wealth, the amount of income/money that comes into the household, pose a unique challenge in the context of educational research.

Although such measurements appear to be rather straightforward conceptually as long one can define a particular monetary standard such as the Euro, asking questions about household income or wealth is plagued by low response rates and questionable accuracy on responses when students are those asked to provide the information (Buchmann, 2002; NCES, 2012).

Additionally, income distributions and the general economic context vary considerably from one country to another. Consequently, comparing wealth measures across countries remains problematic. Typically, such measures as found in both IEA and OECD studies are used in relation to students' performance within each country rather than using country means to explain international variation. Such an approach makes use of each country's economic context without necessarily drawing conclusions about the validity of cross-country comparisons of country means and the relationship of country means of wealth and student performance.

Composite SES Index

Research can be conducted using any one or all three of the typical SES components. This is a central issue in conducting research: Should the component measures be included separately or as a composite indicator (Bradley & Corwyn, 2002)? This is an especially important consideration in international research given that these components are each contextually based and, consequently, "SES indicators are likely to perform differently across cultural groups" (Bradley & Corwyn, 2002, p. 373). Such concerns may in part explain the different approaches taken by the IEA and OECD studies: IEA studies have gathered SES component information but have not created a composite SES index, while the PISA ESCS index is one of the hallmarks of OECD studies. Others have emphasized, however, that combining SES components into a single index yields greater reliability and is more robust (Ganzeboom, De Graaf, and Treiman, 1992; NCES, 2012).

One of the debates about how to conceptualize and interpret the concept of SES in any study centers around what type of capital the index represents. Is the status that is represented by the composite index primarily class-based (meaning *economic* status) or a function of relative social status? Coleman et al.'s (1966) investigation of education made use of three types of capital: economic, human, and social. Bradley and Corwyn (2002) assert that far too few SES indicators have incorporated all three of these. However, all have been present in one form or another in the IEA and OECD studies: economic (books in the home, list of possessions);

human (parent's education); and social (language spoken in the home, number of people living in the home).

The Role of SES in International Assessments

Large-scale assessment studies always demonstrate a large amount of variation among students' scores. The essential question for researchers, educators, and policy makers is what factor(s) account(s) for the observed variance? Is it strictly a function of how well the students have mastered the materials taught in the classroom; or is the variation a function of differences in students' natural abilities; or is the students' home environment a determinant of such differences? Popular understanding of Coleman et al.'s (1966) research left many educators feeling rather irrelevant given the emphasis placed on the role of students' home environment in their school learning. From another perspective, including this in analyses of student learning represents a statistical control for the initial differences among students as they enter and encounter learning opportunities in school (Buchmann, 2002).

Ample evidence existed before Coleman et al.'s (1966) study that family background plays an important role in student learning and performance. Contemporaneously, the FIMS study revealed a strong relationship of family background (SES) with respect to student achievement, and it was not restricted to the United States alone. Yet it was the Coleman report that sparked the often rather heated debate about the relative importance of the influence of family background and schooling on performance.

This debate underscores the importance of maintaining an appropriate understanding of what any SES index or component does and does not represent. Despite the construction mechanics of any such component or index, these must not be considered or interpreted as latent variables such as intelligence or mental ability. Rather, these are proxies for the more proximal factors that *are* related to student learning.

Proxies developed to measure wealth typically include home possessions, home structural characteristics, and so on. They are supposed to approximate long-term wealth – family purchasing power rather than income. In using this asset index internationally, however, it is important to remember that the assets in the index do not necessarily have the same relationship to wealth and status in all countries. Nonetheless, the low nonresponse rates for these types of questions provide a compelling rationale for their continued use (Buchmann, 2002).

Hauser (1994) cautions that a standard set of racial/ethnic and socioeconomic variables cannot serve as all-purpose statistical controls for family background, however well measured. Sound measurements of the social and economic background of children are difficult to acquire. Even after substantial investment in data collection, coding, and data management, it may not capture the major components of family background, broadly construed. Hauser and Featherman (1976) cite examples from studies of the educational attainment of brothers and sisters that even a complete and carefully measured set of social background characteristics would not account for more than about half the resemblance between siblings.

Researchers tend to use father's occupation status to construct family background measures, though this is beginning to expand to also include mother's occupation since it has a stronger relationship to the education outcomes of children, and women are, in today's world, more prevalent in the workforce. Mother's education and father's education, used separately or together depending on the cultural context, have also been used due to their relationship to education outcomes (Buchmann, 2002).

The debate surrounding whether to rely on the father or the mother in measuring family background highlights the importance of context: in some cultures, mothers spend more time with their children, and in other cultures, mother's and father's education is related, suggesting that people tend to gravitate toward like-educated when selecting partners. This leads us into the next section about using SES measures in cross-country comparison of education.

Family background variables are used to control for the relationship of such demographics and student learning so as to examine school effects in cross-country comparisons. The challenge in choosing such variables is to identify culturally relevant elements for different countries (Buchmann, 2002). Buchmann rightly cautions that in international comparative work comparison of the distribution of outcomes (such as achievement scores) across societies requires understanding of the social conditions in the country to truly understand these distribution differences. Further, international comparative analyses must be able to account for differences in educational goals between and within countries, which is what the inclusion of OTL measures attempts to do. As we mentioned earlier and will continue to reiterate throughout the rest of this chapter, this is why neither SES nor OTL should be discussed in isolation from one another.

Fuller and Clarke (1994) observe that the "aggregate influence of schooling in developing countries has probably been overstated due to the under-specification of student background factors … the greatest weakness here is the lack of social class measures that are culturally relevant to the particular society or community being studied. If imprecise SES indicators from the West are simply imported and error terms contain unmeasured elements of family background that are highly correlated with school quality, achievement effects will be mistakenly attributed to school factors" (p. 136).

Nevertheless, for large-scale international assessment studies, the data are not collected by observations but by questionnaires, internationally adopted, that would be administered to primary and secondary school students. Furthermore the questions are about their families not themselves. The challenge is to find evidence to support the comparability claim. To construct an indicator that is measurement-invariant across countries requires variables that would have common meanings across cultural settings.

Over the past hundred years, education studies typically tend to be done within countries. In a meta-analysis of the relationship between SES and student achievement, White (1982) found and reviewed 143 studies from 1910 to 1977. Likewise, Sirin (2005) included 59 articles from 1990 to 2000. Most of the included studies were from regional data collected within the United States.

Since FIMS, every international large-scale assessment study is a testament to the same. Buchmann (2002) summarized the different international studies of SES and education (both attainment and achievement), laying out the outcome measured, the data source, the country/countries in the study, how family background/SES was measured, and the results (see Buchmann, 2002, pp. 156–163). Recent research has continued to demonstrate the importance of family background to achievement in international assessments (Baker, Goesling, & LeTendre, 2002; Chudgar & Luschei, 2009; Montt, 2011; Schmidt, Burroughs, Zoido, & Houang, 2015; Woessman, 2004).

In virtually every research study that includes both SES and student performance variables, the two are strongly correlated. Particularly with respect to this phenomenon, it is essential to recall the proverb that correlation does not indicate causation; that is, one must realize such a relationship does not "explain" student performance differences. Nonetheless, the variation of the strength of this relationship across various countries becomes a subject of interest. Designing the SES component indicators for use in a comparative study, the usefulness of these in the mind of many relies on the extent of invariance of the developed scale. Despite satisfying this constraint, however, it remains a possibility that the causal mechanism at work could be through different family background variables in different countries.

Country wealth is a popular variable to be used to relating SES and student performance. For example, Heyneman and Loxley (1983) examined the relative strength of the relationship from country to country to determine whether the Coleman Report (1966) findings generalized from the US context to other countries. They had a particular interest in comparing students in low per capita income countries to high per capita income countries at the primary level. For this analysis, they used IEA country level data from twenty-nine countries. One of their conclusions was that the SES relationship to achievement varies by a country's per capita income. Industrialized and third world countries are not the same; therefore, inputs that affect achievement are different. They concluded, in contrast to Coleman et al., that school was more important than the home background environment in determining achievement in third world countries; a "fact" that was often considered a given, particularly in World Bank circles in the 1980s (Riddell, 1989).

Riddel (1989), however, suggested that Heyneman-Loxley's conclusion is simply highlighting the inability to accurately measure and detect SES differentiation in third world countries rather than proving that these countries are more homogenous in terms of the within-country SES distribution.

Other studies have suggested that the lack of school effects in industrial countries may be accounted for by the lack of variance in school resources in these countries. In addition, some studies have pointed to a possible threshold effect: A certain GDP level leads to the Heyneman-Loxley (HL) school effect conclusion yet another level yields the Coleman report effects (Gamoran & Long, 2007). More recently, Nonoyama-Tarumi and Willms (2010) used PISA to explore the relationship between the risk of low literacy with low SES and low school resources for countries at various GDP levels. They found that while low SES is associated with greater risk for low literacy in high GDP countries, the nature of this observed relationship is curvilinear. Consequently, they found that low literacy has a greater risk association with low SES family background than attending a poor resourced school irrespective of a country's gross domestic product (GDP).

Other studies have used international data to suggest that economic growth is related to student achievement in the aggregate, as well as to student knowledge across the distribution (Hanushek & Woessman, 2007). In other words, the economic prosperity of a country as a whole is influenced by the educational achievement of average, high-performing and lower-performing students alike.

Carnoy and Rothstein (2013), in reviewing the complexity of international comparisons, concluded that there is a test score gap between socioeconomically advantaged and disadvantaged students in every country. Although the size of the gap varies somewhat from country to country, countries' gaps are more similar to each other than they are different. Countries' average scores are affected by the relative numbers of advantaged and disadvantaged students in their schools. Adjusting for differences in countries' social class composition can change their relative rankings. The United States, for example, has relatively more disadvantaged students than is the case for typical comparison countries. Interestingly, if average scores were adjusted so that each country had a similar social class distribution composition, US scores would appear to be higher than conventionally reported and the gap with top-scoring countries, while still present, would be smaller.

They also concluded that trends in test scores over time vary more by social class in some countries than in others. In the United States, there have been striking gains over the previous decade for disadvantaged students, but not for the more advantaged. In some countries, students perform relatively better on some international tests than others, even though the tests purport to assess the same academic subject area. Explanations for this phenomenon are not immediately obvious but may reflect technical deficiencies in assessments, differences in item types and item emphasis (multiple choice vs. constructed response), differential alignment between assessments and the country's curriculum, variation in student population definitions, or some combination of these and other as yet unidentified explanations.

SES Measures in IEA and PISA Studies

IEA Studies

The IEA founding fathers were researchers with backgrounds in psychology, sociology, mental measurement, education, and the academic disciplines. From the beginning, they were interested in finding the determinants of student performance, which they considered would be a constellation of individual motivation, interest, and background together with classroom context and instruction. The models that informed and guided these early studies were expansive and multi-level, including system-level variables, classroom/teacher variables, and student variables (Foshay, Thorndike, Hotyat, Pidgeon, & Walker, 1962; Keeves & Lietz, 2011). At the heart of all of these conceptual models was the conviction that instruction in schools was an important research focus for any comparative study of student achievement. This was formalized in the curriculum model formulated by Travers and Westbury (1989), which was expanded and elaborated for the 1995 TIMSS (see Chapter 2; Schmidt & Cogan, 1996; Schmidt, Jorde, et al., 1996).

Nonetheless, the early studies took great care to collect information about students' home background. The Pilot Study collected information about fifteen student background factors: birth date, place of birth, sex, number of siblings, language spoken in the home (was it the same as the language of the test?), home location (three categories: city of 20,000–100,000, 2,000–20,000, or under 2,000), the number of years the student had been in school, whether or not the student had attended kindergarten, size of the student's class (six categories: less than 10; 10–20; 21–20; ... 61 or more), mother's and father's education, interest of the parents in education, mother's and father's occupation, and a non-verbal intelligence test score.

The occupation data employed the International Labor Organization (ILO) codes that were available. However, in part because of the constraints of data entry and data analysis – "only two columns on the punch cards could be allowed for occupational status" – Anderson developed a nine-category coding for use in the study with one additional code for nonresponse. One column was for the nine-category coding and the second was to indicate the type of occupation, whether it was "scientific" or "technical" in character (Anderson, 1967).

In commenting on the challenge of collecting all this background information for all students in each of the Pilot countries, Thorndike later observed that "[p]ressures and sensitivities differ from country to country, so that in some it is possible to get information about education and in others it is possible to get information about occupation, but it is rarely possible to get both. Furthermore, the differences in educational structure in different countries make it difficult to establish classifications that will be comparable from country to country." (Thorndike, 1962, p. 27). Consequently, the SES information gathered was not complete and, most likely, not comparable.

After surprising themselves as well as many others upon successfully conducting the Pilot Study, the IEA founding fathers began planning the first, more rigorous international comparison, which became the First

International Mathematics Study (FIMS). From the beginning, they had rejected outright the idea that achievement was essentially hereditary or primarily a function of the student's home environment but rather a more complex endeavor that was greatly influenced by the schools and the curriculum students studied.

This idea shaped the development and use of both OTL and SES measures. The main indicator of SES used in FIMS was father's occupation, although both mother's and father's occupation information was collected (Anderson, 1967). Other SES-related variables were mother's and father's education, and place of residence (meaning rural, farm, size of town or city). The different variables were presented using descriptive statistics (means and standard deviations per country) in tables together with correlations with achievement (Thorndike, 1967a, 1967b). Father's education and occupation were both used to measure relationships with achievement, student plans for further education, interest in mathematics, amount of homework, and program level (academic or general to look at student sorting). Analyzed separately was residence via rural-urban with achievement correlations (Bloom et al., 1967; Husén, 1967b). FIMS definitely improved on instrumentation and methodology yet the SES indices had many problems. The measure for parent education was obtained simply by the number of years of schooling, which did not take into account the rather large differences in education systems. In addition, the questions about mother's occupation only asked whether and how much she worked (Buchmann, 2002).

In the second mathematics study (SIMS), the SES indicators relied on parents' educational attainment and parents' occupation as identified by student questionnaires (Garden & Livingstone, 1989; Robitaille & Garden, 1989; Travers, Oldham, & Livingstone, 1989). SIMS improved on the FIMS measures yet did not code education according to the ISCED categories, relying instead on more general descriptive levels of schooling such as "elementary school," "some secondary school," etc. This prevented the linking of the SIMS data to other data sets. Nonetheless, SIMS was the first to include questions about both parent's occupations on the student questionnaires (Buchmann, 2002).

In SIMS, SES was called the "'Home Status' measure" (Garden & Livingstone, 1989, p. 28), and was a composite of father's occupation – using occupational-category scales at the national level that were nationally appropriate, but recoded to an international scale with four main sublevels to delineate status – and both mother's and father's education. Mother's occupation did not end up being used since no classification of in-home work was available (Garden & Livingstone, 1989; Wagemaker & Knight, 1986).

Occupation information was obtained from the students by asking them to provide a generic name for their parents' position as well as a detailed description of the work done by the parent. For mothers this question also asked students to classify whether their mother worked part-time or full-time and in or out of home. However, there was enough confusion that there was a high nonresponse rate for these questions (Wagemaker & Knight, 1986). The student survey also asked for highest level of education completed by each parent/guardian, but the data were not completely consistent since some national centers, such as Sweden and Scotland, who removed "no schooling" and "primary schooling" as options due to preconceived notions of how often they would be selected, altered the levels. In addition, the concept of tertiary education was not necessarily relevant in all countries (Wagemaker & Knight, 1986).

The SIMS analyses found that student achievement was "related both directly and indirectly to socioeconomic-status variables. Individually, performance on tests was influenced by factors in their homes which enhance learning. But there were also effects which related to the general affluence and aspiration levels of the local communities. The quality of the physical environment of a school, its resources, and its teachers may, in some systems, reflect the socioeconomic status of the school community, as may the educational motivation and aspirations of its students." (Garden & Livingstone, 1989, p. 28).

In reflecting on what had been learned in SIMS, Travers (2011) contrasted the emphasis of the Coleman Report (1966) on student background and socioeconomic status with that in SIMS, which clearly documented *"that curriculum and instruction do in fact make a difference in student achievement"* (Travers, 2011, p. 90).

The TIMSS-95 Conceptual Framework presented in Chapter 2 (see Figure 2.3) illustrates the relationship of student background and SES. The SES constructs are considered to be different sorts of capital, resources students bring with them that can affect the quality and efficacy of their learning in schools. Table 8.1 lists items related to these constructs that appeared on the student questionnaire in TIMSS-95 and how these constructs were operationalized.

Both TIMSS-95 and the following TRENDS in International Mathematics and Science Studies measured SES by parent education and a home possessions index obtained through the student questionnaire for eighth-grade students. The questionnaire for the younger, Population

Table 8.1 *TIMSS-95 student background items*

	Student characteristics		
	Background	Household economic capital	Household cultural capital
Student's Age	X		
Student's Gender	X		
Student's birth country	X		
Student's parents' birth country	X		
In your home, do you have ...?			
study desk/table		X	
dictionary		X	
calculator		X	
computer		X	
<Country-specific wealth item 1>		X	
<Country-specific wealth item 2>		X	
<Country-specific wealth item ...>		X	
<Country-specific wealth item 12>		X	
Parents' Education Level			X
Language of test spoken at home			X
Number of people living in the home			X
Number of books in the home			X
Parental academic press			X
Parental press in mathematics/science			X

1 students (i.e. fourth grade for most countries) did not include questions about parent education since those answers would not be as reliable (Buchmann, 2002; Schmidt & Cogan, 1996). At this point, IEA studies ceased asking students about parents' occupation for the same reason; considerable concern had developed about even eighth-grade students' ability to reliably report about their parents' occupation (Buchmann, 2002). Table 8.1 shows that only a few home possessions items were collected across all countries; each country completed the list of twelve home possessions with appropriate items for their country's context.

PISA Studies

Table 8.2 compares the SES-related items collected in various administrations of the TIMSS-95/TRENDS studies as well as in OECD's PISA. The student population for PISA is fifteen-year-olds, students that are

Table 8.2 SES related items from TIMSS-95/TRENDS and PISA student background questionnaires

		TIMSS-95 & TRENDS (8th Grade)						PISA (15-year olds)						
		1995	1999	2003	2007	2011	2015	2000	2003	2006	2009	2012	2015
Number of Participating Countries		46	38	47	50	42	38	43	41	57	75	65	72
SES Components	Father's Education	x	x	x	x	x	x	x	x	x	x	x	x
	Mother's Education	x	x	x	x	x	x		x	x	x	x	x
	Father's Occupation							x	x	x	x	x	x
	Mother's Occupation							x	x	x	x	x	x
	Father's Employment Status							x	x	x	x	x	
	Mother's Employment Status							x	x	x	x	x	
	Home Possessions List (includes number of books in the home)	x	x	x	x	x	x	x	x	x	x	x	x
Minority and Residential Status	Language Spoken in the Home	x	x	x	x	x	x	x	x	x	x	x	x
	Born in Country? (student, mother, father)	x	x	x	x	x	x	x	x	x	x	x	x
	If no, student's age at migration to country	x	x	x	x	x	x	x	x	x	x	x	x

typically two years older than the eighth-grade students defining the IEA focus. Consequently, questions about parents' occupation reappear in the PISA student background questionnaire.

Unlike the IEA, which included SES-related items but did not provide an SES composite or scaled index, PISA gathers information about students' social, economic, and cultural status and constructs a single index (ESCS) to capture SES. This index is a combination of characteristics of the student's family, specifically capturing parent's education, parent's occupation, and home possessions. More specifically, the ESCS index includes the highest parent occupation status (i.e. mother's or father's), the highest parent education level in years (i.e. mother's or father's), and several indices indicating the home wealth. Table 8.3 identifies these items and the various wealth indices to which they are assigned.

The ESCS index is derived from a Principal Component Analysis. The index itself is then scaled to have a mean of zero across all OECD countries, with a standard deviation of one (Causa & Chapuis, 2009; OECD, 2014b). "The PISA ESCS index is intended to capture a range of aspects of a student's family and home background. It was explicitly created in a comparative perspective by PISA experts with the goal of minimizing potential biases arising as a result of cross-country heterogeneity" (Causa & Chapuis, 2009, p. 8). Table 8.4 displays component loadings and loading ranks, and reliability for the ESCS composite for a selection of PISA 2012 participating countries. In the table the OECD countries are listed first followed by a selection of partner countries. The full list is in the PISA-2012 report (OECD, 2014b). Table 8.4 is provided simply to give a sense of the consistency of the loadings as well as the range of the reliabilities estimated for the ESCS index across countries.

The Relationship between SES and OTL

Consider the following thought experiment: If we could ensure that "all things are equal" and students in the same classroom are all provided the same learning opportunities, would we expect equal performance across all students on an assessment of their learning? Most would probably agree that this would not be the likely outcome. There are many reasons why students perform at a given level on any assessment, reasons both proximal and distal. Reasons range from the condition of the testing environment, the well-being of the student on the day of taking the assessment, what the student ate at breakfast, the proctor misreading the instructions, the family having recently immigrated from a foreign country, the student having

Table 8.3 *Student wealth items included in the PISA-2012 SES index (ESCS)*

	Wealth	Cultural possessions	Home ed. resources	Home possessions
In your home, do you have:				
A room of your own	X			X
A link to the Internet	X			X
A <DVD> player	X			X
<Country-specific wealth item 1>	X			X
<Country-specific wealth item 2>	X			X
<Country-specific wealth item 3>	X			X
Classical literature		X		X
Books of poetry		X		X
Works of art		X		X
A desk to study at			X	X
A quiet place to study			X	X
A computer you can use for school work			X	X
Educational software			X	X
Books to help with your school work			X	X
Technical reference books			X	X
A dictionary			X	X
A dishwasher				X
How many of these are there at your home?				
Cellular phones	X			X
Televisions	X			X
Computers	X			X
Cars	X			X
Rooms with a bath or shower	X			X
How many books are there in your home?				X

Table 8.4 Component loadings, loading ranks, and reliability for the ESCS composite from PISA-2012

	Highest occupational level of parents (HISEI)	Highest education level of parents (PARED)	Number of home possessions (HOMEPOS)	HISEI	PARED	HOMEPOS	Reliability
Australia	0.78	0.78	0.67	1	1	3	0.57
Belgium	0.84	0.81	0.70	1	2	3	0.69
Canada	0.79	0.79	0.66	1	1	3	0.60
France	0.80	0.78	0.70	1	2	3	0.62
Germany	0.84	0.81	0.64	1	2	3	0.65
Greece	0.85	0.84	0.70	1	2	3	0.72
Ireland	0.81	0.81	0.68	1	1	3	0.65
Israel	0.82	0.82	0.67	1	1	3	0.65
Italy	0.84	0.81	0.68	1	2	3	0.68
Japan	0.76	0.77	0.66	2	1	3	0.55
Korea	0.78	0.79	0.73	2	1	3	0.64
Mexico	0.86	0.86	0.81	1	1	3	0.79
Netherlands	0.80	0.78	0.72	1	2	3	0.64
Spain	0.84	0.83	0.67	1	2	3	0.69
Sweden	0.81	0.77	0.65	1	2	3	0.60
United Kingdom	0.78	0.75	0.72	1	2	3	0.59
United States	0.83	0.82	0.73	1	2	3	0.70
Argentina	0.83	0.82	0.75	1	2	3	0.69
Brazil	0.83	0.83	0.78	1	1	3	0.73
Taiwan	0.83	0.82	0.72	1	2	3	0.69
Hong Kong-China	0.85	0.83	0.78	1	2	3	0.76
Indonesia	0.81	0.82	0.79	2	1	3	0.73
Malaysia	0.83	0.76	0.78	1	3	2	0.70
Peru	0.84	0.84	0.79	1	1	3	0.76
Russia	0.80	0.78	0.72	1	2	3	0.63
Singapore	0.85	0.84	0.73	1	2	3	0.73
Thailand	0.86	0.87	0.82	2	1	3	0.76
Uruguay	0.86	0.84	0.78	1	2	3	0.76
Viet Nam	0.82	0.82	0.79	1	1	3	0.74

access to a tutor, guidance from attentive parents, and being homeless, among others. In other words, many of the possible reasons are related to the family background and/or socioeconomic status of the family. Scholars have long debated the relative roles of schooling versus the home/community environment and family background in student learning and performance. Yet this is an artificial division between what the student brings and what the school offers in educating the student.

To study the relative relationships, we need to identify and quantify the proper independent variables. In the broadest sense, the opportunity to learn (OTL) indices discussed in Chapters 6 and 7 capture some sense of instructional time and instructional content covered. However, the education available to a student (OTL) is often a product of social, political, and economic resources as well as the distribution (and re-distribution) of those resources. Education provides social mobility pathways and SES is often used as the factor to examine social inequality in educational attainment. Other school characteristics that may affect student performance include teacher characteristics, pedagogy, and physical characteristics of the school, as well as the backgrounds of other students attending the same school or class. Thus, SES and OTL are inextricably entwined in educational research.

The effect of the structure of the school is often ignored. Students attend classes contained (nested) in schools. There is ample evidence that schools differ in the average SES of their students and that this average school SES is positively related to student performance *and* the OTL provided to them. While SES characterizes what students bring to school, these positive relationships have important implications to the country's educational policies. By the same token, if we were to characterize the SES effect on student performance, we should utilize indices that combine the effects from different levels of the structure of school. Dumay and Dupriez (2007) noted that there is a very close relationship between the process of schooling and the composition of classrooms.

From the TRENDS-2003 dataset, they selected four education systems – Belgium (Flemish), England, The Netherlands, and the United States – that had large between-school/class variance. (Note that class and school variance cannot be separated as only one class per school was sampled.) The results suggest that the joint effect of group composition and class processes, like class climate, explains a significant part of the between-school/classroom variance. The implication is that disadvantaged students tend to be grouped into classes and schools with weaker instruction, materials, etc. This form of implicit tracking illustrates how SES may be related to the opportunities to learn students receive in school.

In general, tracking can be understood as the allocation of students at the same grade in the same school to different classrooms with very different content coverage (Schmidt & McKnight, 2012). It is difficult, and possibly downright impossible, to disentangle the precise contributions of family background, instructional content, and other school-related policies. The reason for this is that the practice of tracking in effect confounds all of these characteristics. For example, as low-SES students are often weaker performers, a system that groups struggling students and offers them more elementary content muddles our ability to determine whether it is peer effects, content, or background factors that most influence student outcomes for disadvantaged students. This is explored more fully in Chapter 9.

In addition, the disputed effect of school resources on student learning might be related to the problem of tracking if low-tracked classrooms receive lower-quality materials. Further, Bradley and Corwyn (2002) theorize that SES impacts teacher attitudes and expectations toward students, and this may in turn impact the in-class experience a student has and as a result impacts educational opportunities. Teachers give low SES children less positive reinforcement; children with low SES backgrounds have fewer cognitive experiences in the home, which reinforces negative stereotypes.

Others suggest that SES and OTL are related based on how parents of varying backgrounds choose to and are encouraged to interact with the schools their children attend, in terms of meeting the teachers, knowing what is happening in the classroom in order to best support it at home, and even having access to the learning materials being taught so as to provide reinforcement (Plowden, 1967a, 1967b). The underlying assumption is that disadvantage in terms of aptitude is related to the income and psychology of the home environment but that may be overcome if parents interact with schools appropriately.

The Effect of SES on OTL and Performance

Despite the many challenges in characterizing student background effects and measuring family background in educational research, international studies continue to rely on SES to capture differences related to family backgrounds. The relationship between SES and performance is not uniform across countries (see Chapter 10). Should that be the case? The range of values in SES indices across countries could simply be related to sampling variation; or to the components of the SES index having different meanings in different countries as they relate to SES.

Even if there are common ways to measure different components of SES, they may yet have different ranges, distributions, and meanings within different countries. Furthermore, the information has to be obtained from the students using paper-and-pencil questionnaires. Such is the dilemma of measuring SES in international research. Although both IEA and OECD have made great strides toward developing SES indicators for use internationally, these are not without their critics. Ultimately, each country must evaluate the usefulness of an SES indicator within its own context.

Comparative reports from large-scale educational assessment studies are incomplete unless they are contextualized with SES and OTL. Horse-race postings always include the jockeys and the handicaps. The same can be said about comparing student performance. Without accounting for SES and OTL differences, overall mean scores are most likely misrepresentations.

Options available to collect information for both OTL and SES are limited by the circumstance of administering large-scale international assessments. Although studies have not adopted a uniform measure of SES, the definitions proposed by the NCES advisory panel for the measurement of SES in the US National Assessment of Education Progress (NAEP) are based on the approaches found in the OECD and IEA studies. Three components are identified: home possessions, parent education, and parent occupation.

In most OECD countries, as Schmidt, Burroughs, Zoido, and Houang (2015) have shown using PISA 2012 data, student SES is related to both OTL and mathematics performance, which raises the important policy question of whether students with different SES backgrounds have equal access to OTL. Such is not the case. Neither SES nor OTL, separately or together, prove to be sufficient in accounting for all the differences in performance. In part, this is because SES not only has a direct effect on performance, as does OTL, it also has a relationship to opportunity to learn as discussed in the previous paragraphs. In this way, SES also has an indirect effect on performance through the direct relationship of OTL to performance. The total SES effect is the sum of the SES indirect and direct effects; the former indicating a schooling SES effect and the latter the effect of home and family background.

This finding indicates the presence of a statistically significant indirect effect related to schooling in all OECD countries (except Norway, which did not provide the necessary data). The average percent of the total SES effect on performance – indicating the total SES inequality – that was

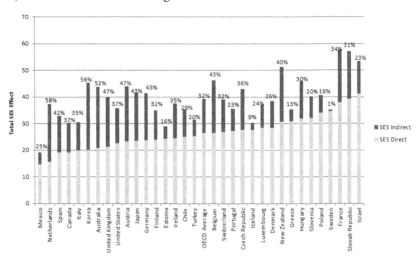

Figure 8.1 Direct and indirect SES inequalities

attributable to the indirect effect through schooling was a third (32 percent). The range of the size of the indirect effect across the thirty-three OECD countries was between 1 and 58 percent.

In other words, the Netherlands more than doubled (58 percent) their total SES effect through the indirect effect created by inequalities in OTL related to SES (See Figure 8.1). Sweden on the other hand, had virtually no relationship between SES and OTL. The total SES inequality could also be subdivided into the parts that were within or between-school, which identified the OECD English speaking countries as having the largest percentages related to within-school inequality, which is likely related to tracking – at least in the United States. While OTL is necessary to learning and thus performance, access to OTL is determined by SES. One can only wonder what the world would be like if the relationship between SES and OTL was eliminated.

Additionally, in discerning the relative contributions of various factors to student performance one cannot ignore simple measurement issues. Mathematical models work best when both dependent and independent variables exhibit a good amount of variation across units such as students and schools. SES as a construct typically has been constructed on the basis of much better measurements than, for example, the quality of instruction – a construct with outstanding face validity yet is exceedingly difficult to operationalize let alone measure – or student motivation, typically measured by a simple three- to five-point scale.

The practical result of this is that important proximal factors for student learning/performance are not measured with anything near the precision of the more distal SES factors yet the SES factors display greater variance across students, which theoretically enables SES to explain a greater amount of variance in the dependent variable. On top of this measurement challenge is the real-world context, which reveals that SES is rather strongly, more strongly in some contexts than others yet never weakly, related to the important proximal factors that drive learning – all of which provides a sound rationale for why SES looms so large in international, as well as country-specific, education studies (Schmidt et al., 2001; Schmidt & McKnight, 2012; Schmidt, Burroughs, Zoido, & Houang, 2015).

Virtually every study of education, whether conducted across nations or within any one, incorporates some measures of students' background, often constructing a composite background index that functions as a measure of SES. The rationale for this was stated at the beginning of the chapter; studies have incontrovertibly demonstrated that a student's family background has an effect on the student's learning. This chapter has been devoted to examining the measurement of SES.

Nonetheless, we must conclude with the reminder that all student background factors, many of which may be and often are combined into a single SES index, are nothing more than proxies for the real factors that affect student learning. Sudden increases in the home's income or the number of cars, TVs, or other possessions or a parent's increase in education level will not dramatically affect or necessarily cause change in the student's learning opportunities, motivation for learning, or background knowledge.

Studies of human development and learning have repeatedly demonstrated these to be the more proximal factors associated with student learning. However strong the association may be between SES, its components, and academic performance, it is imperative that educators, policy makers, and, most especially, parents remember and maintain focus on those factors that truly and more directly affect learning.

The reasons for including SES measures in internationally comparative education studies are compelling. Interpreting the results with respect to the most proximal factors that actually affect student learning is not straightforward; it requires substantial knowledge about the social and educational context as well as how the proximal factors for learning have their impact.

PART III

The Lessons Learned from International Assessments of Mathematics and Science

CHAPTER 9

Pitfalls and Challenges

Having extensive data on the education systems of a large number of countries provides a wealth of useful and important information. This is the role that international educational assessments such as PISA and TIMSS (including TIMSS-95 and TRENDS) have increasingly played, especially over the last twenty years. These data have been influential in the research and policy communities but also in the political context by influencing governmental decisions related to education policy.

Given the nature and design of these studies, there are inherent limitations and potential pitfalls. These are often lost in the dialogue around the results and, as a consequence, the conclusions drawn can be very misleading. In this chapter we explore some of those pitfalls and shortcomings. One such pitfall follows from the scenario, by now quite familiar, that takes place every three or four years as the PISA or IEA results are announced with great fanfare in front of the press. Word rapidly spreads about which country is number one in the rankings. This then leads to the inevitable pitfall – the fatally flawed next question asked by the press, on behalf of the public, but also asked by policy makers and government officials alike: "What are they doing to achieve such great results?"

Seeking the answer to that question has spawned an education tourist enterprise that arranges excursions to the number one-performing country for educators and policy people to find "the answer" by visiting classrooms and interviewing education leaders in that country. That description is not hyperbole and, although it may sound silly to some, it actually represents what has happened. People are searching for answers. They want to know why and how a country gets to the "top." This is, of course, of particular interest to those who are not number one.

Many others take a more studious approach to such questions and study the structure of the top country's educational system, its policies, and the typical instructional practices employed by the teachers as well as their academic and pedagogical preparation. But even this approach to the study of the number one country is a potential pitfall.

Over the years, different countries have been number one. Some of those that have enjoyed that status in mathematics include Japan, Singapore, Finland, Korea, and Shanghai (China). Given how often the number one status changes, it seems that looking for the answer to the "why" question in this way is probably not the best approach.

One example of the problem with this approach relates to Finland. It achieved its number one status among OECD countries in mathematics in the 2003 and 2006 PISA studies but lost that status in 2009 to Korea only to fall behind four more countries in 2012, falling even further in 2015. Even in 2003 and 2006 it was not number one overall, as Hong Kong and Taiwan[1], non-OECD members, held that distinction. In fact, its ranking was lower in each study if the non-OECD participants were included. Further confusion exists when looking at the TRENDS results in 2011, where Finland's performance is more like the international average and, in fact, the performance is not statistically significantly different from the United States, a country not typically viewed as a top-achieving country. The question that arises from this example is; so what does it mean to be number one?

All of this seems somewhat obvious but we labored the point because it is a reality that governments and others do fall into this trap of looking for solutions from a top-performing country. To international comparative education scholars, this approach represents sheer folly. But for others, this seems a quite reasonable approach. For them, fancy regression analyses do not really tell the story. They want to see in a holistic fashion what that country is doing that "works."

There is one suggestion that might make such an approach more reasonable. Rather than focusing on a single country, look at the set of countries that distinguish themselves apart from the rest of the distribution. Here there is less variability in terms of which countries are included over time and over studies. Second, it is not what is idiosyncratic to a specific country that leads to insights for improving performance – as these things are likely to be too culturally and country specific to transfer elsewhere – rather it is what one finds in common across the set of top-performing countries that could be helpful for educational reform efforts elsewhere. If they are in common, they are more likely to be further generalizable to yet other countries. At least there is a greater chance that what is learned in this way might be more useful.

---

[1] We use Taiwan here and throughout the text to refer to Chinese Taipei.

Pitfalls and Challenges 185

This was the approach taken in the TIMSS-95 curriculum analysis toward discovering the key principles of a good curriculum and corresponding curriculum standards. Using the mathematics performance results at eighth grade for the five top-performing countries and studying their intended and implemented curriculum led to the discovery of the three principles of coherence, focus, and rigor, which have been very influential in curricular reform around the world. This is discussed in Chapter 7. A similar analysis in science led to the same principles.

Tables 9.1 and 9.2 list the countries whose performance in mathematics and science, respectively, led to their being ranked among the top five performers according to the total score in studies spanning more than fifty years. IEA studies usually include more than one student population, in which case a country could be among the top five performers on any one of the populations studied. SIMS did not provide a total score but ranked countries on different content areas.

The tables show the large number of countries both in mathematics and in science that achieved top-five ranking on at least one assessment for one population within the sixteen studies (excluding the IEA Pilot Study). Japan stands out in each of mathematics and science, finishing in the top five on at least one assessment in fourteen science studies and ten mathematics studies, as do Korea, Hong Kong, Singapore, and Taiwan. We included these two tables to reinforce the notion that identifying top-performing countries on some assessments for some populations of students leads to twenty-four countries in mathematics and twenty-seven in science. The set of participating countries in which a country might achieve top-five level performance varies across the studies, as do the populations within studies. Consequently, the rankings have different frames of reference and are not directly comparable. What is notable, however, is that some countries consistently perform among the top-five countries, no matter the competitors or the nature of the test or the population (grade levels) of the students tested. This reinforces the idea of studying sets of countries instead of isolated occurrences for answers to the question of "why." The ultimate answer to the question of which countries did best is, it depends.

Beyond the Cross-Sectional Design

The question underlying the above discussion is, "why do certain countries do better than others?" At the heart of this question lies the fundamental

Table 9.1 *Countries ranked in the top five in at least one international mathematics assessment*

Country	FIMS	SIMS	TIMSS1995	TIMSS1999	TIMSS2003	TIMSS2007	TIMSS2011	TIMSS2015	PISA2000	PISA2003	PISA2006	PISA2009	PISA2012	PISA2015
Australia	●	●												
Belgium	●	○	○		○									
Canada			○											
Denmark		●												
Finland	●	●							●	●	●	●		
France	●	●	●											
Germany	○													
Hong Kong		●	●	●	●	●	●	●		●	●	●	●	●
Hungary		●												
Iceland		●												
Israel	●													
Japan	●	●	●	●	●	●	●	●	●					●
Kazakhstan						●								
Korea		●	●	●	●	●	●	●		●	●	●	●	
Liechtenstein											●			
Macau-China													●	●
Netherlands	●	●	●							●	●			
New Zealand		●									●			
Russia			●											
Singapore			●	●	●	●	●	●				●	●	●
Sweden		●	●											
Switzerland		●												
Taiwan				●	●	●	●	●				●	●	●
United Kingdom	○	○												

Note. For SIMS, England & Wales and New Zealand had tied ranks.

● = Country-level participation in the assessment
○ = Only regions or cities participated in the assessment and reported
  = Country did not participate in the study

issue of causality. Given the nature and design of cross-sectional studies, as most international educational assessments are, there are serious limitations with respect to inferring causality. This problem was recognized some forty years ago by Torsten Husén (1979).

> A cross-sectional survey only provides information about students *after* they have been taught, not only by the teacher in charge of a particular subject during a given school year, but by several other teachers during previous school years as well. The lack of information about student status prior to

Table 9.2 *Countries ranked in the top five in at least one international science assessment*

Country	FISS	SISS	TIMSS1995	TIMSS1999	TIMSS2003	TIMSS2007	TIMSS2011	TIMSS2015	PISA2000	PISA2003	PISA2006	PISA2009	PISA2012	PISA2015
Australia	•													
Austria			•											
Belgium	o													
Canada		•	•								•			
Czech Republic			•											
Denmark			•											
Estonia					•						•		•	•
Finland	•	•					•		•	•	•	•	•	•
Germany	o													
Hong Kong		•			•	•				•	•	•	•	
Hungary	•	•	•											
Iceland			•											
Japan	•	•	•	•	•	•	•	•	•	•	•	•	•	•
Korea		•	•	•	•	•	•	•	•	•	•		•	
Liechtenstein											•			
Netherlands	•	•	•											
New Zealand	•													
Norway			•											
Poland		•												
Russia		•				•	•	•						
Scotland	•													
Singapore		•	•	•	•	•	•	•				•	•	•
Slovenia		•					•							
Sweden	•	•	•											
Taiwan				•	•	•	•	•				•		•
United Kingdom	o	o			o	o					•			
United States	•		•											

Note. For PISA 2006, Estonia and Japan had tied ranks.

• = Country-level participation in the assessment
o = Only regions or cities participated in the assessment and reported
 = Country did not participate in the study

being instructed by their present teacher hardly permits valid conclusions about how much competence in a given year is accounted for by the teacher during the year … Above all, the cross-sectional design makes it very difficult to disentangle the influence of school factors from the background factors that the student carries with him to school (pp. 383–384).

He continues by suggesting a study design that actually foreshadowed what was done in SIMS.

> Since IEA wanted to assess what schools did to the children in various national systems of education, the most rational research strategy would have been to *follow* the students at various levels over at least 1 school year and to measure the *gains* achieved during that period among pupils taught by the same teacher (Husén, 1979, p. 383). Highly sophisticated methods of multivariate statistical analysis cannot substitute for the lack of longitudinal information over time (p. 384).

SIMS addressed Husén's concerns by designing an eight-country longitudinal sub-study with students tested in both the fall and spring of the school year. The focus of the study was on student *learning* not what students knew. As such it was a study of the relationship of various instructional practices and OTL to the *gain* in student performance over the school year. The study was conducted at the grade level within which the majority of thirteen-year-olds were enrolled; for most countries that was eighth grade.

The longitudinal study focused, as all IEA studies do, on opportunity to learn but also focused on "what teachers do as they teach mathematics" (Burstein, 1993, p.3). Both of these foci were important to the ultimate goal of the research, which was to study how much mathematics students learn and how that is accomplished. This was a break in IEA tradition and to this date is the only longitudinal international assessment of mathematics to be done, fulfilling the dream of Husén. Volume III of the SIMS reports contains sections that reflect the focus of the resulting analyses: "The Identification and Description of Student Growth in Mathematics Achievement" (p. 59) and "Understanding Patterns of Student Growth" (p. 101).

Being able to go beyond the limitations of a cross-sectional design, as SIMS did, paid strong dividends as the discussion was able to be moved from status (what is known) to gain (what has been learned). This was not only true in terms of the league tables now ranking countries on growth as opposed to achievement status but also in strengthening the results from the relational analyses linking gain or growth to home/background characteristics and schooling in the form of OTL, teacher characteristics, and classroom processes (see Burstein, 1993).

This suggests that the international studies of achievement together with the various audiences of the results of the study would greatly benefit from using a longitudinal design. The problem and the answer as to why SIMS has been the only IEA study done in this way is the cost of such a study and the difficulty in following the students over the course of a year. While some discussion occurred in the planning stage leading up to the third mathematics study (TIMSS-95) about taking this approach, the idea of a longitudinal study was quickly dismissed as TIMSS-95 became much more complex.

This brings us to still another approach for addressing the limitations inherent in cross-sectional studies. If the assumption is made that classrooms, in the two adjacent grades within the same school, are similar with respect not only to their demographic characteristics but also with respect to their schooling except for the additional year of instruction that is the focus of the study, then the performance of the lower grade can serve as a pre-measure to the performance of the upper grade. This then provides a measure of growth or gain for the upper grade that is the focal population for the study. It mimics a longitudinal study but where the unit of analysis is the classroom or school (or country) rather than the individual student (Wiley & Wolfe, 1992).

This cohort longitudinal design was used in the TIMSS-95 study, a design approximating the SIMS longitudinal study. It did not require two time points for data collection as adjacent grades were tested in the spring of the same school year. As described in Chapter 2, the two populations were age-defined grade levels for nine- and thirteen-year-olds. For most countries the cohort longitudinal design sampled grades three and four, and grades seven and eight as the definition of the Population 1 and Population 2 samples, respectively. Population 3, defined as students in their final year of secondary education, employed the traditional IEA cross-sectional design (see Chapter 2).

At the country level, using the performance of the lower grade as a pre-measure to define the gain at the focal upper grade is reasonable unless the country as a whole has recently experienced a major shift in the demographics between the two adjacent cohorts. At the classroom level this seems reasonable since both grades are in the same school. The same caveat exists about large shifts in the two populations but now at the school level.

Taking advantage of these assumptions, various analyses were done to model the growth and how it is related to measures of home and family background as well as measures of schooling, especially OTL. These analyses are reported in *Why Schools Matter*, illustrating the strong relationship of schooling as measured by opportunity to learn and related to the measures of learning (residual gain or gain) controlling for socioeconomic status (SES).

The one complexity in performing these analyses in TIMSS-95 occurred when the country tracked students into classes with different content coverage at the same grade level. This occurred primarily at the lower-secondary level – grades 7 and 8. Here matching the seventh-grade class with its particular content coverage with the appropriate eighth-grade class had to take into account the tracking system used in that school. Not only does content tracking of the sort discussed throughout this book create such difficulties but so does ability tracking. Several countries had such complexities but the country with the largest challenge in this regard was the United States, which at that point in time practiced pervasive tracking, particularly in mathematics.

The issue is different for the PISA studies in spite of both being based on a cross-sectional design. PISA is not intended to be a curriculum-focused study as are the IEA studies. That difference is important in that the TIMSS-95 goal was to relate curriculum coverage and classroom instruction during the academic year to student performance on a curriculum-defined assessment near the end of the academic year.

For PISA the goal is to determine how those same factors plus others over the whole of the students' schooling up to and including the grade at which the student turns fifteen is related to cumulative performance on a problem-solving literacy test. Here the factors influencing performance are by definition cumulative in nature since "literacy" is not focused on at a specific grade. The relevant school-based content coverage and instructional activities are presented over the years of schooling but especially in the lower-secondary grades.

Given the assumption that performance on a literacy-based test for fifteen-year-olds is also cumulative with an initial zero point (school entry), the logic of the relationship between those cumulative school-related experiences and performance provides stronger support for a causal inference.

Partitioning the Variance in International Studies

Performance and background variables in both IEA and PISA studies vary across and within countries. More importantly, within countries they vary between schools, between classrooms within schools, and between students within classrooms and schools. Such is the nature of school organizations – and of individual differences. The characterization and the modeling of that variation, both overall and at the various levels of schooling, is the focus of analyses related to understanding schooling.

Consequently, in international studies the most important question to investigate is not only how much overall variation there is but also *where* that variation is located in the hierarchical structure of schooling and how large it is. This brings us to the examination of variance components and their importance.

Given the hierarchical organization of schooling in virtually all countries, the overall variation across students within a country, defined by mean-deviating the individual student score by the overall country mean, can be divided and allocated to represent the portion of the total variation that corresponds to each level of the structure of schooling. This is important not merely for statistical reasons but because each of the sources of variation (i.e. organizational level) have different interpretations with concomitant policies and practice implications.

The top-down, hierarchical, multilevel organization of schooling – similar among countries that have participated in PISA and TIMSS (including TIMSS-95 and TRENDS) – may be summarized as follows:[2]

I. Major regional designation (typically geographic – e.g. province, state, etc.)
II. Local designation (e.g. town, district, city, county, etc.)
III. Schools
IV. Tracks within schools (where present)
V. Classrooms, which includes teachers (within tracks)
VI. Students within classrooms

This hierarchical multilevel organization highlights the nested nature of students and schools that must be taken into account in research design and analyses of educational data. In the following discussion, levels I and II are merged for simplicity but also because most countries do not include them in the sampling design other than as stratification variables; having done that, the differences across strata can be recast as variance contributors. For this reason we include them in what follows. This leaves five levels in the typical hierarchical (nested) structure of schooling but only

---

[2] A more precise definition of the six levels would be:
    I. Major regional (typically geographic) designation (e.g. province, state, etc.)
    II. Local designation (e.g. town, district, city, county, etc.) within major regional designations
    III. Schools within the defined regional entities
    IV. Tracks within schools (where present) within defined regional entities
    V. Classrooms, which includes teachers within tracks within schools within defined regional entities
    VI. Students within classrooms within tracks within schools within defined regional entities

four if there is no formal tracking in the country (unfortunately tracking can also be hidden). We first examine the apportionment of variation excluding tracking, given that many countries do not acknowledge tracking, at least in grades K–8.

The overall variation within a country is represented by the total variance of student scores. It is an important indicator for the country alongside the mean, as it speaks to the issues of inequality. Let Y be a student's score and the following means are defined for each of the four levels identified as follows:

Mean of all students: μ
Level 1 – Region/District mean (I and II): μ_a
Level 2 – School mean (III): μ_b
Level 3 – Classroom mean (V): μ_c
Level 4 – Individual student's score (VI): Y

We can decompose the student score, Y, to represent the four components of the simplified structure of schooling from the following tautology:

$$\begin{aligned} Y &= Y + \mu - \mu = \mu + (Y - \mu) \\ &= \mu + (Y - \mu_c) + (\mu_c - \mu) \\ &= \mu + (Y - \mu_c) + (\mu_c - \mu_b) + (\mu_b - \mu) \\ &= \mu + (Y - \mu_c) + (\mu_c - \mu_b) + (\mu_b - \mu_a) + (\mu_a - \mu) \end{aligned} \qquad 1.$$

We add subscripts to represent all the students in a country and at the same time indicate each level with a different notation to denote the component of the score:

$$\begin{aligned} Y_{ijkl} = \ &\mu \quad \rightarrow \text{GrandMean} \\ &+ (\mu_{a_i} - \mu) \rightarrow a_i \rightarrow \text{Effect for Region/District}_i \\ &+ (\mu_{b_{ij}} - \mu_{a_i}) \rightarrow b_{ij} \rightarrow \text{Effect for School}_j \text{ within Region/District}_i \\ &+ (\mu_{c_{ijk}} - \mu_{b_{ij}}) \rightarrow c_{ijk} \rightarrow \text{Effect for Classroom}_k \text{ within School}_j \\ &\qquad\qquad\qquad\qquad\qquad\quad \text{within Region/District}_i \\ &+ (Y_{ijkl} - \mu_c) \rightarrow e_{ijkl} \rightarrow \text{Individual Difference for Students}_l \text{ within} \\ &\qquad\qquad\qquad\qquad\qquad\quad \text{Classroom}_k \text{ within School}_j \text{ within} \\ &\qquad\qquad\qquad\qquad\qquad\quad \text{Region/District}_i \end{aligned} \qquad 2.$$

Resulting in Equation 3:

$$Y_{ijkl} = \mu + a_i + b_{ij} + c_{ijk} + e_{ijkl} \qquad 3.$$

This leads to the partitioning of the student variance of Y into the variance for each of the four components:

$$\sigma_y^2 = \sigma_a^2 + \sigma_b^2 + \sigma_c^2 + \sigma_e^2 \qquad 4.$$

where, σ_a^2 is the regional variance,
σ_b^2 is the average of variances of school means within region/districts,
σ_c^2 is the average of variances of class means within schools, and
σ_e^2 is the average of variances of students within classrooms

Equation 4 shows the partitioning of the overall student variance into its components — each reflecting variation in student performance at a different level in the typical multilevel, nested organization of schooling. This allows for the estimation of the relative contribution of each component to the total variance. It is usually done as a percentage with statements such as the between-school variance component (σ_b^2) accounts for 20 percent of the total variance. An estimate of σ_b^2 in Equation 4 specifies the actual size of the component. This can also be modeled in terms of its relationship to other variables defined at that same level.

The challenge that is present in estimating the variance components and in interpreting them correctly results from differences between the reality of the structure of schooling in a country and the associated sample design actually employed by the country in the study. The requirement for correctly estimating the three main levels identified in Equation 4 (considering region as a fixed stratification variable) is that a random sample must be employed at each of the three levels. In other words, there must be a random sample of each of schools, classrooms within schools, and students within classrooms (or all the students within the classroom) in order to estimate the three variance components. As mentioned previously, the stratum differences can be used to get a measure of variation. Additionally, the sample must include at least two units at each of the levels, e.g. schools, classrooms, and students, otherwise there is no variation in the sample by which to estimate the corresponding variance component in the school system.

The consequence of not meeting these two requirements is biased estimates for those components that must represent more than one level of the school structure due to a lack of sample variation at the lower level. For example, in the absence of two classrooms within schools, the only component that can be estimated is at the school level but that estimate represents both classroom and school (σ_b^2 and σ_c^2) variation combined. This creates a problem in interpretation since what is estimated as school variation is confounded (not separately estimable) with differences among classrooms since the sampling design did not include a random sample of two classrooms from each school. In other words, the two components

σ_b^2 and σ_c^2 are inextricably intertwined in the estimation procedure. This impacts the ability to draw clear implications for policy.

For example, in the original TIMSS-95, the greatest proportion of the total student variance in Japan at eighth grade was estimated to be accounted for by individual differences within classrooms and measurement error (89 percent) and not by school-level differences (11 percent). While in the Netherlands the opposite was true with 55 percent of the variance accounted for by school-level differences. In both, only one classroom was sampled in each school; and as a result the estimate of the school variance component includes both school and classroom variation combined, not allowing the differentiation of variation across schools in the Netherlands from that present across classrooms within schools. The implications are profound, as in Japan the means by which students are sorted either into schools or into classrooms within schools has little effect on the variability related to performance, while in the Netherlands the sorting mechanism of who attends which school or who is assigned to which classroom accounts for over half of the variation in student scores. This difference between the two countries most likely reflects tracking at either or both levels, which is practiced to a great extent in the Netherlands and very little in Japan.

The above situation can be illustrated as follows: If only one classroom is selected from each school, the estimation model cannot separate components for school and classrooms. This yields Equation 5 for estimating the available components.

$$\begin{aligned}\hat{\sigma}_y^2 &= \hat{\sigma}_a^2 + \widehat{\left(\sigma_b^2 + \sigma_c^2\right)} + \hat{\sigma}_e^2 \\ &= \hat{\sigma}_a^2 + \hat{\sigma}_{b+c}^2 + \hat{\sigma}_e^2\end{aligned} \qquad 5.$$

Only two countries in TIMSS-95 included a sample of two classrooms from each school. As a result, for all participating countries, except Australia and the United States, the estimate of the school variance component ($\hat{\sigma}_{b+c}^2$) includes the variance component for classrooms as well as schools similar to Japan and the Netherlands example, as discussed earlier (see also Figure 9.1).

For Australia, if we add the two separate variance components for classrooms and schools, we get a ratio of between-school to within-school variation very similar to other countries in Figure 9.1 such as Germany and Hong Kong. If Australia had not included two classrooms per school in their sampling design they would have attributed 46 percent of the total variation to school effects, which would be very misleading because only 5–6 percent is really due to schools, as most of the variation is

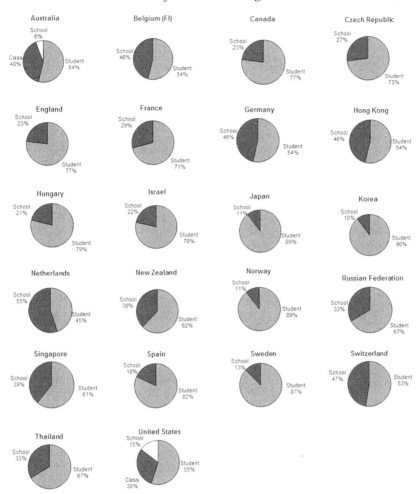

Figure 9.1 Sources of variance for the eighth grade International Mathematics score.
Note. Republished with permission of Springer Science and Business Media BV, from "Facing the Consequences: Using TIMSS for a Closer Look at U.S. Mathematics and Science Education," by W. H. Schmidt, C. C. McKnight, L. S. Cogan, P. M. Jakwerth, & R. T. Houang, 1999, p. 174; permission conveyed through Copyright Clearance Center, Inc.

characteristic of classroom differences. The policy and practice implications for classroom differences are very different from school-level issues.

To further illustrate the confusion that can occur in studies of schooling, we include three other examples. If only one school is drawn from each region/district, we can only estimate a single component for both combined (and not separately) region/district and school.

$$\hat{\sigma}_y^2 = \left(\widehat{\sigma_a^2 + \sigma_b^2}\right) + \hat{\sigma}_c^2 + \hat{\sigma}_e^2 \qquad 6.$$

Furthermore, if only one school is drawn from each region/district, and one classroom is selected within each school, we can only separately estimate and distinguish the variance components for within classrooms from the others, e.g.

$$\begin{aligned}\hat{\sigma}_y^2 &= \left(\widehat{\sigma_a^2 + \sigma_b^2 + \sigma_c^2}\right) + \hat{\sigma}_e^2 \\ &= \hat{\sigma}_{a+b+c}^2 + \hat{\sigma}_e^2\end{aligned} \qquad 7.$$

The final example refers directly to the type of sampling used in PISA, where the sampling design includes schools with no classrooms, as students are sampled directly from the population of fifteen-year-olds at that school. Here the estimate of the within-school variance component does not describe individual differences with respect to performance on a mathematical literacy test alone but also includes classroom and grade variation as well. Furthermore, the classroom and grade components (not separately estimable) also include differences in teacher quality and differences in the content coverage as well as other factors. The resulting estimation model for the PISA sample design is:

$$\begin{aligned}\hat{\sigma}_y^2 &= \hat{\sigma}_a^2 + \hat{\sigma}_b^2 + \left(\widehat{\sigma_c^2 + \sigma_e^2}\right) \\ &= \hat{\sigma}_a^2 + \hat{\sigma}_b^2 + \hat{\sigma}_{c+e}^2\end{aligned} \qquad 8.$$

Due to the differences in the sampling designs between PISA and TIMSS (including the original TIMSS-95 and TRENDS), the variance component estimates for those countries who have participated in both types of studies are not comparable. Estimates of the within-school variance component reflect both individual differences and classroom variation in PISA, while in TIMSS classroom variation is not separately estimable from school variation.

Returning to Figure 9.1, we raise the issue of tracking, which is of particular relevance to the United States. The estimated variance components in TIMSS-95 suggested that 55 percent of the total student variance was at the individual student level within classrooms with the bulk of the remaining variance at the classroom level within schools. The school-level estimate was 30 percent of the total. It is here that one must consider the prevalence of tracking in 1995 and what effect that has on the estimate of the classroom component.

To estimate a tracking variance component separate from classrooms it would be necessary to randomly sample at least two classrooms within each track. This has not been done in any IEA study to date; but in TIMSS-95, as described in the next section of this chapter, additional information was available that facilitated the development of an estimate for the tracking variance component.

As a foreshadowing of the next section, schools with no tracking (see Table 9.8) on the one hand, yielded an estimate of the classroom variance component that accounted for 25 percent of the total variance, with 16 percent at the school level – very similar to the estimates in Figure 9.1. On the other hand, for schools with tracking, the tracking component was estimated at 40 percent with only 8 percent at the classroom level – a very different set of implications from the unbiased estimate of classroom variation (see Table 9.9). These results can only be considered as an approximate estimate of the variance component related to tracking and separately from classroom variation within tracks, given the limited data that was available. To what extent such tracking exists in other countries is not well understood since it is not a required variable for developing the official country sampling design. Nonetheless, it is extremely important, as this has profound policy implications. For this reason, in the next section we present a detailed study of the effect of tracking on the variance components related to performance in the United States to help understand this important but mostly ignored aspect of schooling.

Confounding OTL with Tracking: An Example from the United States

Given the centrality of OTL as a measure of schooling in international studies, the impact of the confounding, as described in the previous section, of tracking with classrooms and consequently with OTL, must be taken seriously. This confounding yields either an over or underestimation of the relationship of schooling to performance. This biasing occurs in IEA studies because only one classroom is typically sampled in a school. In the TIMSS-95 US sample, however, two classrooms were sampled within each school yet the sampling design did not include a tracking stratum even though such tracking was known to be widespread. We use the US data to show the implications of tracking in the only country where the available information made it even somewhat possible to do so. Not being a requirement of the sampling rules of the studies, the extent of tracking across countries is not well known.

Three different kinds of mechanisms have been identified that may produce effects of tracking on student performance: social, institutional, and instructional (Gamoran & Berends, 1987; Lucas, 1999). The focus here is on the instructional effects – specifically those related to OTL – in order to show how tracking effects are confounded with OTL, leading to an underestimation of the role of OTL on performance.

Course Differentiation

As part of the US TIMSS-95 within-school sampling design, schools were required to list all of their seventh- and eighth-grade mathematics classes along with the class titles and the list of the students enrolled in each class. This was used to draw the sample but it also provided complete tracking information for all of the sampled schools. Using this information, it was possible to specify the within-school course-offering structures (Cogan, Schmidt, & Wiley, 2001). It is that within-school course-offering structure that defines tracks for our purpose here. More than twenty-five different patterns of school course offerings were identified in the sample, based on six types of classes: remedial, regular, pre-algebra, enriched algebra, and geometry. Each of the six types of classes defines a track in the sense of providing different content opportunities to learn (OTL) mathematics. The actual number of tracks was probably much larger.[3]

As very few remedial, enriched, or geometry classes appeared in the actual TIMSS-95 sample, a reduced number of course types (regular, pre-algebra, and algebra) were employed in the analyses of the US data to define tracks and to more fully examine their effects. Based on other information in the study that provided more extensive knowledge of course content, enriched courses were recoded as pre-algebra, remedial as regular, and geometry as algebra (there were only a few such cases).

An analysis of the school tracking forms revealed two types of schools. The first type offered a single type of mathematics course to all eighth-grade students. The second type of school offered multiple courses or

---

[3] By far the most prevalent title for a grade eight mathematics class was simply "math" or "mathematics." A number of variations on this title were also observed – many of which incorporated the notion of tracking students according to ability. Examples of this approach included "average mathematics," "basic mathematics", "advanced mathematics," "gifted" or "high "mathematics, "LD mathematics," "remedial mathematics," and "resource mathematics" among others. The only other commonly occurring class titles that did not contain either "math" or "mathematics" were "arithmetic," "pre-algebra" and "algebra." Courses in some schools carried such unique titles that their content focus and relation to a progressively unfolding mathematics curriculum was unclear (Cogan et al, 2001).

tracks into which different students were assigned. Approximately 27 percent of US eighth-grade students attended a school in which there was only one mathematics course available to them (Cogan, Schmidt, & Wiley, 2001). Although these schools might group students into different sections based on ability, there was no explicit content differentiation among the sections. For these non-tracked schools, content coverage should be the same for all eighth-grade students in the same school, at least in terms of official school or district policy, even if there are multiple sections of the same course offered.

The other type of school attended by the vast majority of eighth-grade students (73 percent) offered two or more different types of mathematics courses or tracks covering different mathematics content for different eighth-grade students. The combinations of tracks offered within a school based on the three course types (which itself is a simplification) are many. For example, the popular impression in the United States that most tracked schools offer the three basic types of courses that include regular mathematics, pre-algebra, and algebra was true for only 30 percent of US eighth-grade students who attended tracked schools. Some schools did offer those three tracks (attended by around one-fourth of all eighth-grade students) but other schools offered different paired combinations of the three types, with the most common being regular mathematics and algebra. This type of school was attended by one quarter (25.2 percent) of the eighth-grade students.

Previously, specific-topic differences were described among the various tracks (Cogan, Schmidt, & Wiley, 2001). Here, however, we employ the International Grade Placement (IGP) index as an indicator of content coverage for the entire year. The metric of the index is grade level (see Chapter 7).

Statistically significant differences were evident in the IGP index across the three types of courses (whether offered within a tracked or non-tracked school): regular mathematics, pre-algebra, and algebra ($p < .0001$).[4] The estimated contrast of the algebra course with the combination of the pre-algebra and regular mathematics courses was statistically significant (see Table 9.3 for the summary statistics). The estimated value indicated an almost one year difference in content coverage (.88) between the algebra and the other two types of courses. This implies that there are substantial differences in content coverage between the different types of courses – at least between algebra and the other two course types.

---

[4] This and all analyses throughout the chapter include school in the design so that any clustering effects are appropriately accounted for.

Table 9.3 *Means, standard deviations, and sample sizes for schools and classes by type of school and class track*

	Schools	IGP		Classes	International scaled score		Students
	N	Mean	Std Dev	N	Mean	Std Dev	N
Non-tracked schools	47	7.28	0.57	85	484	93	1822
Regular		7.27	0.43	70	481	90	1509
Pre-Algebra		6.70	0.65	3	522	96	81
Algebra		7.51	1.02	12	495	110	232
Tracked Schools	134	7.45	0.84	258	505	90	5124
Regular		7.11	0.63	132	469	79	2506
Pre-Algebra		7.28	0.76	67	517	85	1356
Algebra		8.20	0.75	59	565	79	1262
Totals	181**			343			6946

Note. Means and standard deviations are weighted. **Two schools were dropped due to lack of information

This issue may also be addressed by taking into account the school structure in which the course occurs. To do this, separate analyses of variance were done on the same IGP index for non-tracked and tracked schools. Overall there were no differences in IGP between tracked and non-tracked schools (p < .09). The means were almost identical – 7.45 vs. 7.28 – which is equivalent to about a two-month difference in content difficulty.

In tracked schools, the algebra track was statistically significantly different from the other two tracks in content rigor (p < .0001). The difference between the pre-algebra and regular mathematics tracks was also significant (p < .02). The estimated contrasts indicated that the algebra track classrooms were covering content slightly over one grade level higher (1.09) than the regular track and almost one grade level (.92) more advanced than the pre-algebra track.

For the non-tracked schools a somewhat different pattern emerges with respect to algebra. The content difficulty of the coverage for classrooms in schools (n = 12) that offer only algebra is not significantly different (p < .14) from the coverage for classrooms (n = 70) that are in schools that offer only regular mathematics. That difference is only about one quarter of a year. However, the average values of the IGP index for the algebra

Pitfalls and Challenges

Table 9.4 *Variation in OTL in non-tracked schools*

Source	Estimated variance	Estimated percentage of variance
Course Type	.019	5
Schools within course type	.066	17
Classrooms within schools	.302	78
Total across all classrooms	.387	

classes offered in non-tracked schools is about three-fourths of a year (p < .03) less rigorous than for the algebra classes offered in tracked schools. On the other hand, the content of the regular mathematics classes in tracked schools is less rigorous by about two months than that of the regular classes in non-tracked schools.

Classroom Variation in Content Coverage

One way to examine how the track structure influences the variation in content coverage across classrooms is to estimate the variance components associated with each level of the school structure. Since OTL was measured at the classroom level, there was no within-classroom variation in OTL. Using the IGP index as a reflection of the complexity of the content coverage, standard statistical algorithms were used to estimate the variance components associated with schools, tracks within schools, and classrooms within tracks within schools for the tracked schools. In the case of the non-tracked schools, variance components were estimated for course type, schools within course type, and classrooms within schools. Tables 9.4 and 9.5 present these results for non-tracked schools and tracked schools, respectively.[5]

Note that non-tracked schools do not all offer the same course type. Some offer only regular mathematics, some only algebra, and very few only pre-algebra. Therefore, some of the variation among non-tracked schools in OTL reflects the fixed differences that stem from the different courses being offered by the different schools, i.e. course type. The variance in

---

[5] Variance components were computed using the maximum likelihood estimation procedure available from the SAS procedure PROC MIXED. Where appropriate, student-level sampling weights were also used.

Table 9.5 *Variation in OTL in tracked schools*

Source	Estimated variance	Estimated percentage of variance
Schools	.214	25
Tracks within schools	.336	40
Classrooms within tracks	.300	35
Total across all classrooms	.851	

OTL as scaled by IGP at the school level reflects variation across schools in which the same course type is offered. This likely results from different interpretations across schools as to what content is part of a course titled Regular Mathematics, for example, or from the use of different textbooks.

The results for the tracked schools (Table 9.5) clearly reflect the impact that tracking has on content coverage. In schools with tracks, 40 percent of the total cross-classroom variation in content coverage was attributable to track differences. Around 25 percent of the variation is due to school differences in terms of course offerings, while around 35 percent of the variation was attributable to cross-classroom within track within school variation, which we hypothesize was mostly reflective of teacher differences in their interpretation of the content or other adjustments made relative to textbooks or students characteristics.

The estimated total between-classrooms within-schools variation for non-tracked schools (.302) is less than half the size of the same component for tracked schools (.336 + .300). This implies that tracking actually increases the variation in content coverage across classrooms. The increased variation occurs primarily because of the track level, as the actual value of the classroom component within tracks is essentially the same for both the tracked and non-tracked schools (.300 and .302).

Track Differences in Achievement

Since tracking starts in the seventh grade and continues through eighth grade, it is desirable to also look at differences in the gain in achievement over the two year period for different track patterns. Given this goal, the best type of data would be longitudinal data on the same students so that the tracking effect at eighth grade could be separated from prior learning.

Table 9.6 *Classroom means for four patterns of mathematics tracks, including their appropriate seventh-grade feeder class, and eighth-grade gain*

Reg 7- Reg 8 Track			Reg 7 – PA8 track			Reg 7 – A8 track			PA7-A8 track		
7th Grade Regular Mean	8th Grade Mean	Gain	7th Grade Regular Mean	8th Grade Mean	Gain	7th Grade Regular Mean	8th Grade Mean	Gain	7th Grade Pre-Algebra Mean	8th Grade Mean	Gain
462	469	7	468	517	48	462	550	88	524	569	45

TIMSS-95 did not provide such longitudinal data, but did provide cohort-longitudinal data.

We paired eighth-grade classrooms in each tracked school with an appropriate seventh-grade feeder classroom from that same school as defined by the school course-offering structure recorded in the tracking form. Four types of paired track patterns were formed. The first pattern was a seventh-grade regular mathematics class leading into an eighth-grade regular mathematics class; a second track pattern was the same seventh-grade course leading into pre-algebra at eighth grade. The other two patterns both end up with eighth-grade algebra. One starts with regular mathematics at seventh grade while the other starts with seventh-grade pre-algebra. These were the dominant patterns available in the data.

Table 9.6 shows the average mathematics scores for these four track patterns. The unit for these analyses was the paired classrooms on which the two measures were available – the mean seventh-grade achievement score and the mean eighth-grade score both averaged over the students in the pair of classrooms and then averaged across all schools with that pattern. For each pattern, the data in the table give three values: the seventh-grade mean of the feeder classroom, the eighth-grade mean and the gain defined as the difference between the two means.

Differences are apparent in the mean achievement across the two seventh-grade tracks and across the three eighth-grade tracks. Seventh-grade classrooms teaching "regular" mathematics scored around 60 points lower than the pre-algebra classrooms. Correspondingly, there is about an 80 to 100-point difference between eighth-grade regular track classrooms and algebra track classrooms, with pre-algebra classrooms falling in between (517). This nearly 100-point difference between the algebra and the regular mathematics track at eighth grade represents about a one standard deviation difference in the TIMSS-95 test score. The difference between the pre-algebra and regular mathematics track was estimated as roughly one-half of a standard deviation.

These analyses were done only on the tracked schools. Previously, we indicated that OTL (IGP) differed only slightly between the tracked and non-tracked schools on average. Similarly, the difference in eighth-grade mean achievement is also small (505 vs. 484) – about two-tenths of a standard deviation difference – but is statistically significant ($p < .0001$). See Table 9.3 for the summary statistics related to eighth-grade achievement. A more careful examination of the pattern indicates an interaction effect. For algebra classes, the 70-point difference in mean achievement between those in tracked schools versus non-tracked schools is significant ($p < .003$). The algebra students in non-tracked schools are presumably a representative sample of the school population, but in a tracked school the students represent an elite group. The differences in mean achievement for the other two types of courses are not significant. Finally, across the non-tracked schools there were no significant differences in eighth-grade achievement for the three different types of courses ($p < .38$).

Classroom Variation in Achievement

The variance components for the achievement scores (see Tables 9.7 and 9.8) are of interest here as a comparison to those reported for OTL. In the tracked schools, about 40 percent of the variation in achievement across students is related to track differences. This estimated track component is very similar in magnitude to the 40 percent estimate for content coverage as defined by the IGP index. For achievement variation, the fourth variance component reflects within-classroom or individual student variability. This component was estimated to be around half of the total variability for both tracked and non-tracked schools. In other countries in TIMSS-95 such as Japan, Korea, Norway, and Sweden the estimated variance component for students within classrooms was around 90 percent – see Figure 9.1 (Schmidt, McKnight, Cogan, Jakwerth, & Houang, 1999, p. 174).

Taking into account track differences reduces the classroom component to less than 10 percent from the 30 percent earlier reported in Schmidt, McKnight, Cogan, Jakwerth, & Houang (1999), where the track component was not estimated. In non-tracked schools, the class component was around one-fourth of the total variation.

Table 9.7 *Variation in mathematics performance in schools having no tracks*

Source	Non-tracked schools	
	Score variance	Score variance (%)
Course type	0[6]	0
Schools within course type	1328	16
Classrooms within schools	2179	25
Students within classrooms	5074	59
Total across all students	8531	

Table 9.8 *Variation in mathematics performance score in schools having tracks*

Source	Tracked schools	
	Score variance	Score variance (%)
Schools	0[5]	0
Tracks within schools	3431	40
Classrooms within tracks	652	8
Students within classrooms	4414	52
Total across all students	8497	

Relationship of Tracking to Achievement

The pattern of achievement differences across classrooms indicates that, on average, the achievement level of a class is related to the track of the class. This is certainly consistent with many other studies and is not particularly surprising. The analyses presented previously in Table 9.5 did not control for the selection bias introduced by the fact that students were not randomly distributed across the different tracks within schools. Differences in achievement levels across tracks could be attributed to the selection bias associated with who is counseled into the various tracks or who self-selects into each

---

[6] The unconstrained estimated variance component was negative, which was then set to zero. The other estimated variance components are conditional on that value. Additional analyses show the variation across this factor to disappear when adjusting for the other levels in the design.

track. How to disentangle those effects from other potential effects such as a curriculum effect related to content coverage (OTL) is difficult. We explore this relationship in several ways using different statistical adjustments in an attempt to understand the nature of the relationship between track as a curriculum issue and student achievement in mathematics.

To explore this issue more fully, a three-level hierarchical linear model was fitted separately for the tracked schools. The conceptual model defining the analysis model and the particular choice of variables follows the framework defined in Figures 2.1, 2.2, and 2.3. The three levels included schools, classrooms nested within schools, and students nested within classrooms within schools. The track designations were included as indicator (dummy) variables at the classroom level. The model also included several covariates at each level in the design. The student-level model included racial/ethnic identity and a composite SES measure. The class-level model included the appropriate seventh-grade score as the pre-test, mean SES, and track. The school-level model included the school-level mean SES, and three variables derived from the school questionnaire, including the percent minority enrollment at the school, the location of the school (rural, suburban, or urban), and the size of the school as measured by the number of eighth-grade students.

Results are presented in Table 9.9. At the individual student level the racial/ethnicity identity of the student as well as the SES of the family were significantly related to their performance on the TIMSS-95 test ($p < .001$). The racial/ethnic estimated coefficients indicated a large negative relationship to achievement associated with being African American or Hispanic American.

After adjusting for the student-level relationships, the estimated class-level model indicated a statistically significant relationship for track even when controlling for the aggregate SES of the class and the mean level performance of the seventh-grade feeder classroom ($p < .0001$). Although not an entirely perfect solution, adjusting for the prior achievement at the class level and SES both at the class and individual level should remove a substantial portion of the likely student selection bias. This makes the estimated track effects less reflective of selection bias and, therefore, more likely to reflect differences in OTL and/or some other social or institutional effects related to track membership – all of which are related to schooling.

The significant track effect is present both in terms of differences between the pre-algebra track and the regular track ($p < .0001$) as well as in achievement differences between the algebra track and the regular

Table 9.9 *The relationship of tracking and other student, classroom, and school variables to eighth-grade mathematics achievement in tracked schools*

Students-level variables	Estimated coefficient	SE
Race: White (D)	2.41	(7.10)
Race: Black (D)	-26.53***	(7.70)
Race: Hispanic (D)	-21.01**	(8.02)
Race: Asian (D)	-2.50	(8.92)
Socioeconomic Status (SES) (C)	-2.85***	(0.44)
Classroom-level variables		
7th grade achievement	1.00*	(0.49)
Mean classroom SES	20.18***	(2.69)
Class type: Algebra (D)	62.19***	(6.29)
Class type: Pre-Algebra (D)	31.05***	(6.97)
School-level variables		
Mean school SES	-4.97	(3.00)
8th grade enrollment	0.01	(0.02)
Minority enrollment (%)	-0.06	(0.11)
Location: Urban (D)	4.40	(6.06)
Location: Rural (D)	19.73**	(6.92)

Note. D denotes dichotomous indicator variables. C denotes centered variable. SE denotes standard errors. * $p < .05$. ** $p < .01$. *** $p < .001$. Appropriate TIMSS sampling weights were used in the analysis.

track ($p < .0001$). The estimated effects indicate about one-third of a standard deviation difference in achievement between each of the three tracks (again, controlling for the other variables in the model). This implies two-thirds of a standard deviation difference in performance between the typical regular track student and his/her counterpart in the algebra track. In the international context, this two-thirds of a standard deviation is not inconsequential, as it is the difference between the mediocre US eighth-grade performance, i.e. at the international mean, and the performance of two of the top achieving TIMSS-95 countries – the Czech Republic and Flemish-speaking Belgium.

Does Content Coverage Mediate the Effects of Tracking?

These analyses demonstrate the effect of tracking on residual achievement gain from seventh to eighth grade after adjusting for likely sources of selection bias. The estimated effect sizes are large and important. The question is, how do these effects occur? Researchers have suggested three different mechanisms, one of which is instructional, which we further define for purposes of this study as the level of demand or rigor of the content coverage (IGP).

From a content point of view, on average there is a major difference between the level of demand associated with the content covered in the algebra track compared to the other two. That difference was estimated to be about one year. This seems reasonable on the face of it since the algebra track is considered to be the equivalent of high school algebra I, which for most US students is taken at ninth grade - one year later than eighth-grade regular mathematics. The difference between pre-algebra and regular mathematics is smaller, and for the most part, not statistically significant.

Additional analyses were done to see if we could further differentiate OTL across the three tracks. With the above concerns in mind, we did a hierarchical linear model analysis in which we placed both track and OTL (IGP) in the same classroom-level model. All of the other variables as described in Table 9.8 were incorporated in the model.

Central to the point of this analysis is whether the estimated track effect is significantly reduced by the inclusion of a measure of content coverage (OTL) in the model. Without control for the background of the students or content coverage the estimated track effect for the algebra track compared to the regular mathematics track was a full standard deviation difference in achievement. After controlling for prior achievement and SES, the estimated effect size for the algebra track compared to the regular mathematics track was about two-thirds of a standard deviation (62.19 – see Table 9.9).

Controlling for OTL – one of the proposed mechanisms by which tracking has its impact – the estimated track effect for algebra is further reduced by about 15 percent to slightly more than one-half of a standard deviation (.51). Instructional content appears to be a mediating factor in how tracking influences academic achievement. The fact that the track effect is not reduced even further by the inclusion of OTL seems surprising, given how inextricably intertwined they are – almost by definition.

One explanation for this might be the limitations of the IGP index as a measure of OTL. Some of the OTL topics such as linear equations that are scaled at the higher level of the IGP scale include a wide range of subtopics representing different levels of content demand, among which the scale does not distinguish. For example, solving a simple linear equation such as $3x = 12$ is coded as the same level of IGP as solving two linear equations with two unknowns or solving quadratic equations.

To address this we divided each of the three tracks into three sub-tracks based on the distribution of the IGP within each track. More specifically, we divided the classrooms in the tracked schools within each track into three groups defined by high, medium, and low content demand. The definitions for each group were different for each track. This creates a nested design with nine levels.

The estimated effect size for track levels with the inclusion of the more refined measure of OTL in the model is now reduced by half from the previous estimate of around 62, as found in Table 9.9, to 28 (s.e. = 14; $p < .046$). This suggests that, after controlling for the selection bias as to who is placed in which track, about one-half of the remaining track effect is a result of its indirect effect on achievement through OTL. Put simply, of the total estimated effect of tracking on performance (about one standard deviation – one year of instruction) about 40 percent is due to differences in student background across the tracks; 30 percent is due to OTL differences across tracks; and 30 percent likely includes other instructional factors such as teacher content knowledge, as well as social and institutional factors as suggested by Gamoran and Berends (1987). If tracking exists within a school system, these analyses demonstrate why it should be treated formally in the sampling design so that its effects can be separately estimated and isolated in order to appropriately inform policy.

CHAPTER 10

What Has Been Learned about the Role of Schooling
The Interplay of SES, OTL, and Performance

In light of the hundreds of millions of dollars that have been spent on OECD and IEA studies over the past sixty years, it seems entirely appropriate, if not completely incumbent on a book about international educational assessments, to address the question posed in the title of this chapter: what have we learned from this enormous examination of schooling conducted across almost 100 countries and regions of the world? We address this question by examining results from the IEA and OECD studies at the center of prior chapters, i.e. FIMS, SIMS, FISS, SISS, TIMSS (both TIMSS-95 and TRENDS), and PISA (primarily the mathematics-focused PISA-2012).

As noted in the preface and by the choice of the book title, our goal in writing this book was to summarize what has been learned from international comparative assessments about the fundamental role that schooling plays in student learning both of mathematics and science. In this chapter we examine the role that schooling and, in particular opportunities to learn, play in student learning in these two essential disciplines. About the same time as the first international study of mathematics, the results of a study done in the United States by the sociologist James Coleman were released, with the subsequent interpretation that home and family background were more important to student learning than schooling.

The debate that ensued is still present some fifty years later. About twenty years ago, the book *Why Schools Matter* drew on analyses of the TIMSS-95 results to make the point that schooling does have a strong role to play in understanding differences in student learning across forty-plus countries and also within those same countries across schools. In this chapter, we again revisit that issue but with a much broader sweep, looking at the results of seventeen studies conducted in around one hundred countries and over a timeframe of sixty years.

What has preceded this chapter has described the historical development and evolution of these studies, who has participated, the methodologies used, and how the studies were designed and executed. This chapter presents what we (the authors) believe has been learned relative to the question raised in the 2001 book title; in what ways *do* schools matter?

More specifically, we focus on the educational policy-relevant insights gleaned from a macro-level view and explore the cross-country results that have emerged, particularly from the more current studies of PISA and TIMSS (both TIMSS-95 and TRENDS). Furthermore, the focus here reflects what has been found that has international implications as opposed to those that are country specific. Such country studies contain results and conclusions that, quite appropriately, address country-specific issues that depend on substantial prior understanding of the national system and, consequently, are not likely to reflect insights of interest to a broader audience. Three notable exceptions, Russia, Germany, and the United States, are addressed, as the effects related to opportunity to learn are broad based and carry international implications. Especially important is that these results have been published and in that way made accessible to those beyond national boundaries.

Before turning to the major findings related to the role of schooling, we first examined the literature to see to what extent secondary data analyses have been completed and published in books, journals, and other types of publications. Restricting our literature search to the 12 data sets available beginning with the 1995 TIMSS, we found a total of 146 articles that have been published in over 50 different publications including major American, European, and Asian journals. The articles are related to multiple issues, including educational policy, mathematics education, science education, economics of education, sociology of education, educational psychology, anthropology, methodological issues including evaluation, and learning and development. Table 10.1 summarizes this review by year and the name of the publication.

To determine the key findings across all seventeen studies, we relied on the official study reports and the associated books, articles, and white papers written by the key researchers who were part of the team that developed the conceptual models, designed the studies and instrumentation, and analyzed the data collected. Every one of these seventeen studies produced a set of such reports, books, or journal articles explicating the results, confirming hypotheses, and raising additional new questions for further research.

Most importantly, we focus on the main theme identified first by the IEA founding fathers, which subsequently has been instantiated in each of

Table 10.1 *Sampling of journal articles using TIMSS/TRENDS and PISA data since 1995*

Year	Journal(s)	# Articles
1995	Educational Evaluation and Policy Analysis	1
1997	Educational Researcher	1
1999	Review of Educational Research	1
2000	Review of Educational Research; Educational Evaluation and Policy Analysis	2
2001	Educational Evaluation and Policy Analysis; Educational Researcher	4
2002	American Educational Research Journal	1
2003	American Educational Research Journal; Economics of Education Review	2
2004	American Educational Research Journal	1
2005	American Educational Research Journal; Educational Evaluation and Policy Analysis; Economics of Education Review; Education Economics; Educational Research and Evaluation	6
2006	International Journal of Educational Research; European Economic Review; Oxford Review of Education	3
2007	International Journal of Science and Mathematics Education; Educational Researcher; Economics of Education Review; School Effectiveness and School Improvement	4
2008	American Educational Research Journal; Review of Educational Research; Educational Research and Evaluation; Studies in Educational Evaluation; Effectiveness and School Improvement	6
2009	American Educational Research Journal; Economics of Education Review; Journal of Education Policy; Educational Research and Evaluation; Compare	5
2010	American Educational Research Journal; Economics of Education Review; International Journal of Science and Mathematics Education; Educational Research for Policy and Practice; African Journal of Research in Mathematics, Science and Technology Education	8
2011	Economics of Education Review; Educational Researcher; European Journal of Education; International Review of Education; Observatoire Francais des Conjonctures Economiques	10
2012	Economics of Education Review; British Educational Research Journal; Educational Researcher; Journal of Comparative Economics; International Review of Education; Asian Economic Journal; Solsko Polje; International Journal of Science and Mathematics Education; ZDM Mathematics Education	11
2013	American Educational Research Journal; Economics of Education Review; PloS one; Intelligence; Revista de Educación; Learning and	18

Table 10.1 (*cont.*)

Year	Journal(s)	# Articles
	Individual Differences; Journal of Development Economics; Educational Research and Evaluation; Computers & Education; International Education Studies	
2014	Economics of Education Review; Journal of Educational and Behavioral Statistics; Educational Research and Evaluation; Anthropologist; International Journal of Science Education; Asia Pacific Education Review; International Journal of Science and Mathematics Education; Frontiers of Education in China; Egitim Ve Bilim; Solsko Polje; Large-scale Assessments in Education	15
2015	Economics of Education Review; Educational Researcher; International Journal of Science Education; Studies in Educational Evaluation; Large-scale Assessments in Education; Computers & Education; SpringerPlus; International Review of Education; KEDI Journal of Educational Policy; American Educational Research Journal; British Journal of Educational Technology; The Asia-Pacific Education Researcher; Asia Pacific Education Review; Learning and Individual Differences	18
2016	Large-scale Assessments in Education; Asia Pacific Education Review; Quality & quantity; Learning and Individual Differences; British journal of sociology of education; International Journal of Science Education; International Electronic Journal of Elementary Education; AERA Open; Economics of Education Review; The Asia-Pacific Education Researcher; American Educational Research Journal; Studies in Educational Evaluation; Educational Studies in Japan; Asia Pacific Journal of Education; Social Psychology of Education; Educational Researcher	23
2017	International Journal of Comparative Sociology; International Journal of Science and Mathematics Education; Journal of Educational Computing Research; Perceptual and motor skills; International Journal of Educational Research; Economics of Education Review	6

the IEA studies and now (starting in 2012) included in PISA: the relationship of schooling – particularly the emphasis on opportunity to learn (OTL) – and family background (SES) to each other and to student performance on assessments defined by academic curricular content, skills, and reasoning, as well as problem-solving competencies defining mathematical or scientific literacy.

What follows are ten major findings in mathematics and science education that have emerged consistently from these seventeen studies. These results are not based on a formal meta-analysis but represent the authors' judgements of which set of findings have been reinforced over the past sixty years and tell the story of schooling. There is no attempt to be exhaustive of all that has been learned; or to reflect what has been reported in IEA and PISA reports or in the 146 journal articles and publications cited in Table 10.1. Neither should the exclusion of some findings be interpreted that we, the authors, deem them unimportant. We believe these ten listed here are central to the goal laid out by the title of this book – to contribute to the world's understanding of how schooling can be improved for all children.

Key Findings

I. *Content coverage (OTL) of both mathematics and science varies substantially across countries and within countries. Within country variation in OTL exists between schools but also within schools both across classrooms and among students.*

This major finding was found in the earliest of the IEA studies and later was found to hold true for all subsequent IEA studies as well as for PISA-2012. The original TIMSS-95 study provided the most detailed evidence, which was extensively documented in the two volumes of *Many Visions, Many Aims* – the first reports of the international curriculum analysis study (one each for mathematics and science) (Schmidt, McKnight, Valverde, Houang, & Wiley, 1997; Schmidt, Raizen, Britton, Bianchi, & Wolfe, 1997). The corresponding US report, *A Splintered Vision,* documented the enormous variation within the United States (Schmidt, McKnight, & Raizen, 1997). Such variation was evident in the intended curriculum (standards documents), the implemented curriculum (teacher- or student-provided content coverage data), and, the potentially 246, implemented curriculum (textbooks). Even for mathematics – the subject matter was selected for the first IEA study because the founding fathers considered it to be similar in substance across countries, thereby making cross-country comparisons more reasonable – what was found were major cross-country differences in both the intended and the implemented curriculum. Also surprising was the substantial variation in OTL within countries even for those that claimed to have uniformity through national standards (See Chapter 4 in *Why Schools Matter,* Schmidt, McKnight, Houang, Wang, Wiley, Cogan, & Wolfe, 2001).

In 2012, for the first time, PISA added measures of content coverage that provided additional support for the finding of large variation in OTL both

What Has Been Learned about the Role of Schooling 215

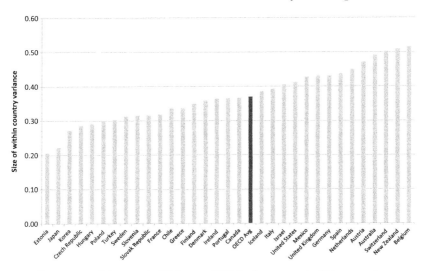

Figure 10.1 Variation in PISA-2012 OTL both across and within countries

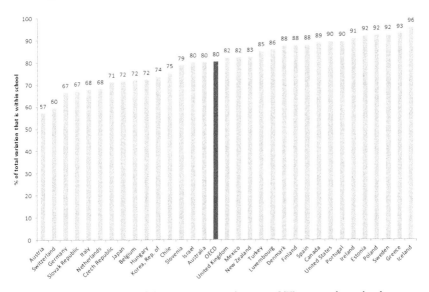

Figure 10.2 Most of the variation in PISA-2012 OTL was within school

across and within countries (see Figure 10.1). The surprising characteristic for most countries is that the percent of variation in OTL attributable to within and not between-school differences was around 70–90 percent (see Figure 10.2). Because of the age-based sampling design – resulting in multiple grades in the sample for most countries – some of this within-school variation is attributable to grade-level differences in OTL (Schmidt, Burroughs, Zoido, & Houang, 2015).

Both PISA and IEA studies evidenced support for the same adage: "School mathematics is not school mathematics" – even for two students in the same school let alone in the same country or in two different countries. Such large variation in OTL suggests the very real possibility that it is related to student performance. This leads to the second major finding.

II. *Variation in content coverage (OTL) for both mathematics and science in its various instantiations (standards, textbooks, teacher instruction) is strongly related to student performance – not only on the typical curriculum-based assessments in mathematics and science (IEA) but also on the literacy assessment focused on mathematics (PISA-2012).*

This, perhaps the most striking and important of the findings, has recently gained even greater stature and relevance after finding the statistically significant relationship of OTL to performance on the PISA mathematics literacy assessment for all of the participating sixty-plus PISA countries (see Table 6.2). The only difference across countries was in the strength of the relationship (Schmidt, Cogan, & Guo, in press).

The TIMSS-95 study found that the variation in each of the intended, implemented, and potentially implemented curriculum bore a strong relationship to performance (Schmidt & Houang, 2007; Schmidt, McKnight, Houang, Wang, Wiley, Cogan, & Wolfe, 2001). Those relationships were found at the country and within-country levels and for both mathematics and science. Although not to the same level of depth, similar findings of the relationship of OTL to performance were found in all of the previous and subsequent IEA studies. Combining the TIMSS-95 results with those from PISA-2012 gives powerful evidence of the relationship of OTL to performance, be it based on a literacy-defined or a curriculum-defined mathematics assessment.

Such overwhelming consistency in over sixty countries in PISA and over forty in TIMSS-95 and in virtually every other IEA study seems to imply the advent of the recognition that schooling as measured by the type and the amount of content coverage (OTL) – a fundamental element of both Carroll's Model of School Learning (Carroll, 1963) and Bloom's Model of Mastery Learning (Bloom, 1974, 1976) – is causally related to student performance. The replication of this relationship under the vastly different conditions represented by the educational systems of this large number and highly varied set of countries with enormous political, economic, demographic, and sociologically related differences would, together with the simple logic inherent in the fundamental idea that students learn that to which they are exposed, provides strong evidence to support a causal inference.

Two challenging questions remain. First, what other conceivable unobserved differences could there be that are not present in the thousands of schools studied, the backgrounds including SES of the millions of children tested, and the almost 100 countries' educational policies and structures that would challenge the causal nature of the observed relationship of content coverage (OTL) to performance? The stability of the relationship across time, studies, countries, and varying conditions is remarkable. Herein lays one of the often unrecognized strengths of international research. The concern in interpreting relationships, not derived from randomized studies, as causal rests on the logical issue of the presence of other factors, not measured, which are related to both OTL and performance (Schneider, Carnoy, Kilpatrick, Schmidt, & Shavelson, 2007). That concern seems small in light of the ubiquity of evidence to the contrary across so many varied educational contexts.

The second challenge some would put forward concerns the absence of a pre-test in most of the IEA and PISA studies. Without such a pre-measure, the relationship of OTL to performance could reflect the reality that higher performance reflects better students; that better students get better OTL (a reality made evident in Chapter 8) and as a result have higher performance. That argument would seem not to challenge the basic relationship of OTL to performance being causal but only to impact the magnitude and structure of the relationship. Furthermore, SIMS (longitudinal), TIMSS-95 (cohort longitudinal), and the Russian longitudinal study from 2011 to 2012 all show that the inclusion of a pre-measure does not eliminate the basic relationship of OTL to performance; it only makes the implied causal structure more complicated. In the end, these cautions serve to remind us how difficult it is to establish causal inference in such studies, yet they do not negate the strong evidence that schooling does in fact matter.

III. *Variation in the amount of exposure students have to applied real-world problems in mathematics (not traditional word problems) in their classroom instruction and in the tests they take is related in a non-linear manner to their performance on a literacy test (PISA-2012).*

Controlling for the relationship of formal school mathematics OTL to performance on the literacy test, the amount of exposure to real-world applications is also positively related to performance but only up to a point.

This non-linear relationship is best approximated by a quadratic function (See Figure 6.2), indicating that at an inflexion point defined as somewhere between "not much" and "some" exposure, the increased exposure to applied mathematics is negatively related to performance. Unfortunately, the limitations in the OTL measurement in PISA-2012 do not allow for any more definitive specification of the inflexion point. Simply put, in most of the countries in which student exposure to real-world problems in their mathematics instruction contributes positively to student performance, this holds only up to a certain point, beyond which the relationship is negative. Further studies are needed to better understand the nature of this relationship. This finding relates only to the PISA literacy assessment.

IV. *Variation in formal school mathematics content coverage (OTL) and variation in the amount of exposure students have to studying mathematics in an instructional context emphasizing applications are both related to literacy performance but in a complex way.*

For PISA the combination of key findings II and III leads to the conclusion that greater exposure to both formal mathematics and exposure to instruction related to real-world applications of mathematics combine in a unique way. Each contributes something unique to the picture of the relationship of schooling to the development of mathematical literacy. For both, more exposure predicts higher performance but only up to a certain point, after which the two work in opposition to each other. This is of interest to most countries, especially with the growing focus on twenty-first-century skills and how what is learned may be applied broadly outside an educational setting. The complex relationship can best be understood in terms of two key issues:

> The issues of learning and teaching are inextricably linked. As soon as one specifies something to be learned, a competency to be developed, a learning objective to be achieved, the question "how shall this be accomplished" immediately follows. This becomes an instructional or pedagogical issue: what is to be done in the classroom to move students in the desired direction, to develop the desired competencies? Responses to this question have historically contrasted a focus on "pure" versus "applied" approaches to mathematics teaching (de Lange, 1996). Instruction focused on "pure" mathematics centers on the study of the academic aspect of mathematics: its definitions, relations, and structure. Upon mastery of these, students may then be asked to apply their knowledge to real world situations that require this type of mathematics knowledge for a solution.
>
> In contrast, the applied approach focuses on connecting mathematics to the real world at all times. The meaning and value of mathematics' rules

and definitions must be illustrated through practical, real world situations suggesting the relevance of a different and a more applied type of OTL in understanding performance on a literacy test. This emphasis on teaching mathematics through applications has been linked with recognizing mathematics as a useful tool in the social sciences and society more generally (de Lange, 2003). If de Lange is correct in his assessment this suggests the potential relevance of such OTL to mathematics literacy.

This contrast between school mathematics and mathematics embedded in real world applications is evident through: the type of representations emphasized for student learning and problem solving (Goldin, 2008), the structure, discussion, and type of problems included in mathematics textbooks (Michalowicz & Howard, 2003), and discussions about the extent to which all students should learn mathematics, the type of mathematics they should learn, and the types of problems that are suitable for them to work on as they learn it (Clements, Keitel, Bishop, Kilpatrick, & Leung, 2013). As Clements and colleagues (2013) noted, there is no clear alignment across all of these issues that provides a simple contrast between traditional, structural, or analytical mathematics and real-world applications and problem solving in characterizing the reform movements in the US and around the world over the past 150 years (Schmidt, Cogan, Guo, in press).

V. *The process of placing K–9 students within the same school and at the same grade level into different classes each of which covers different mathematics or science content – the practice of tracking – is related both to OTL and SES and must be accounted for in the sampling design as well as in the analyses in order to avoid biased estimates of the relational coefficients, particularly that for OTL to performance.*

Tracking is practiced in some countries, especially at the lower and upper-secondary level. This has been the case at the lower-secondary level in the United States, at least in the most recent past, especially when the TIMSS-95 data were being collected. There has been some evidence that the practice of tracking is lessening, but not fully eliminated, due to the influence of the Common Core State Standards of Mathematics (CCSSM). The extent to which this is practiced in other countries is less well understood due to political sensitivities and the fact that tracking, resulting in different content coverage, can be the result of other practices such as ability tracking, which results in the watering down of common courses by creating different sections of a course for different students. In this way, the OTL differences are hidden.

As shown in Chapter 9, there is a strong relationship between OTL, SES, and tracking. The analyses presented in that chapter show that one-third of what would be identified as the effect of tracking on performance can be accounted for by differences in OTL. In addition, membership in the different track levels is related to SES. There is also a residual effect of tracking, after accounting for OTL and the selection bias, that is likely due at least in part to the "stigma" of tracking and its effect on motivation (see Gamoran & Berends, 1987).

In countries where tracking within schools is present, the sampling design must include a specification for the track effect. Without this the various relational coefficients will likely be biased, especially that between OTL and tracking with performance and with each other. The policy implications are different, for example, when tracking still has a significant relationship to performance after accounting for OTL as contrasted with the situation where the tracking effect essentially disappears when controlling for OTL. Similarly, in estimating the between-class, within-track relationship to performance, without a specification of track in the sampling frame, these two very different sources of variation are confounded. Consequently, the very different policy-relevant implications cannot be discerned.

VI. *The background of students including gender and socio-economic status are statistically significantly related to their performance both on the curriculum-based assessments in mathematics and science (IEA) as well as on the literacy based assessments (PISA).*

SES was found to be statistically significantly related to performance in every country of every study; however, its definition varied across studies. What are the reasons for such a universal finding? Four are offered:

(1) The effects of SES on performance are cumulative, so they correlate with the pretest (whether measured or not) and therefore predict performance very well;
(2) Within the year of the study, there is some contribution to performance gain due to motivation, home support, etc.;
(3) SES correlates with OTL, as illustrated in Figure 10.3 and as discussed in Chapter 8,which is related to performance;
(4) SES correlates with teacher and school qualities, which in turn are correlated with performance.

Whatever the reason or reasons, the simple correlation of SES and performance indicates that schooling is not distributed equally and, if nothing is done, gaps will only increase.

The IEA studies as described in Chapter 8 did not develop an index of SES, while the PISA studies did (ESCS). The variables included in IEA

What Has Been Learned about the Role of Schooling

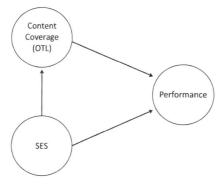

Figure 10.3 Model of SES inequality

studies, however, are essentially the same as those used in the development of the PISA ESCS index with some minor differences. The variables included in each study are summarized in Table 8.2.

To examine whether the two studies' measures of SES are comparable we took advantage of the fact that sixteen countries participated in all three of the following studies – TIMSS-95, TRENDS-2011, and PISA-2012.[1] The student-level correlations of SES with performance in each of those three studies are given in Table 10.2 for each of the sixteen countries. The results are amazingly consistent within countries across the three studies. Overall, the correlation of the country correlations of SES with performance ranged from .63 to .80. All of this is to say that, no matter which definition of SES was used, the estimated relationship of that measure of SES to performance was very consistent across the three studies within each country.

Overall, there are no consistent patterns related to gender differences in performance between the PISA-2015 literacy assessment for fifteen-year-olds and the TIMSS-2015 fourth grade and eighth grade curriculum-based assessments.

On average, fifteen-year-old boys performed better on the PISA-2015 mathematical literacy assessment (5 points) than did girls; however, this statistically significant difference in performance favoring boys was true in only twenty-eight of the PISA participating countries. In some countries, the amount by which boys outperformed girls was as high as 27 points. In

---

[1] A composite SES index was constructed using three key components that were similarly measured in each of the three studies: (1) number of books in the home, (2) number of home possessions, and (3) highest level of parent education.

Table 10.2 *Student-level correlations of composite SES measures with performance within each country for TIMSS-95, TRENDS-2011, and PISA-2012*

Country	TIMSS-95	TRENDS-2011	PISA-2012
Australia	0.380	0.480	0.351
Hong Kong	0.220	0.328	0.267
Hungary	0.420	0.534	0.479
Israel	0.280	0.422	0.412
Japan	.	0.355	0.320
Korea	0.370	0.423	0.321
Lithuania	0.340	0.427	0.370
New Zealand	0.330	0.479	0.423
Norway	0.250	0.389	0.268
Romania	0.320	0.499	0.436
Russia	0.280	0.314	0.343
Singapore	0.270	0.369	0.380
Slovenia	0.320	0.408	0.389
Sweden	0.280	0.397	0.325
Thailand	0.240	0.326	0.312
US	0.360	0.415	0.387
Average	0.311	0.410	0.361
Correlations of Country Correlations		*TRENDS-2011*	*PISA-2012*
	TIMSS-95	0.801	0.632
	TRENDS-2011		0.747

contrast, girls outperformed boys in a statistically significant way in eight countries – Finland being the only OECD country in which this was true. In seventeen OECD countries there were no statistically significant differences in PISA mathematical literacy between fifteen-year-old boys and girls.

At the fourth grade in TRENDS-2015 boys had a statistically significantly higher average performance on the curriculum-based assessment in eighteen out of the forty-nine participating countries. Girls outperformed boys in eight countries, leaving twenty-three countries with no statistically significant gender differences. At eighth grade, the pattern was evenly distributed, with twenty-six out of the thirty-nine countries showing no statistically significant gender differences and in only six countries did boys outperform girls by an average of 9 points, while girls outperformed boys in seven countries by an average of 14 points.

Given the hierarchical nature of mathematics as found in the K–12 curriculum, it would appear that any gender differences may well be related to the level of mathematics being taught and tested at the different grade levels represented across IEA and PISA studies. The stereotype suggesting boys are better at mathematics than girls does not have strong support. Only 33 percent of the potential 158 gender differences from PISA-2015 and TIMSS-2015 for fourth and eighth grade favor boys and 61 percent of those were from performance related to PISA mathematical literacy for fifteen-year-olds.

VII. *Socio-economic status is statistically significantly related to opportunity to learn mathematics resulting in different content coverage for different students. This creates SES-related inequalities in performance that go beyond those resulting from the home and family background of the students to include an additional indirect component related to inequalities in schooling (OTL).*

Virtually every major nation of the world stipulates in its official standards documents the goal of equality of opportunity for *all* students. Increasingly, nations are concerned with the growing gap in income and wealth between the top 25 percent and the bottom 25 percent. They view education as the means to give all students, especially those in the bottom 25 percent of the SES distribution, a chance at a better life. Yet the reality of a PISA analysis shows that the playing field is not level and that schooling actually increases the total SES inequality effect. The total SES effect is defined as the sum of the direct effect of SES on performance (controlling for the direct effect of OTL) and the indirect effect that comes from the relationship of SES to OTL and from the fact that OTL is directly related to performance (see Figure 10.3)

Put simply, the results of statistically fitting the model represented by Figure 10.3 indicates that students with different SES backgrounds do not experience the same exposure to important mathematics and, as a result, fall further behind in their performance (see Schmidt, Burroughs, Zoido, & Houang, 2015). Among the thirty-three participating OECD countries the average SES-related effect of schooling accounted for about one-third of the total SES effect on performance. This ranged from virtually 0 percent for Sweden to over 50 percent for Australia, Korea, and the Netherlands (Figure 8.1 from Chapter 8 is reproduced here in Figure 10.4 for the convenience of the reader).

Another way to characterize SES-related inequalities is to examine the "relationship of between-school inequalities in OTL to between-school inequalities in SES" (Schmidt, Burroughs, Zoido, & Houang, 2015, p. 4). The schools within each country were divided into "SES quartiles relative to that country's average school SES" (p. 4). In each of the top and the

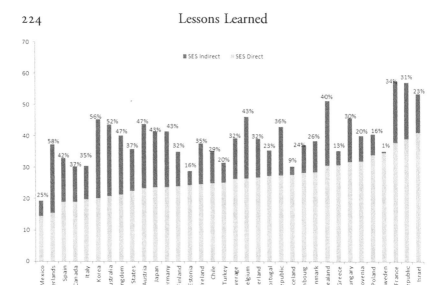

Figure 10.4   Direct and indirect SES inequalities

lowest quartile the country average OTL and performance were calculated as was the difference (gap) between the two quartiles for each of OTL and performance. The country-level analysis found a statistically significant relationship of the average OTL gap to the average performance gap, accounting for almost half of the variance in the performance gap (see Figure 10.5).

VIII. *The pursuit of excellence in terms of performance in mathematics can ironically produce greater inequality.*

The fact that the total SES inequality in mathematics performance has as one of its two components the indirect effect of SES, which is mediated through the relationship of OTL to performance (see Figure 10.3), produces the situation in which a country focused on improving its national standards toward achieving greater excellence in performance could also be inadvertently increasing its SES-related inequalities. If this same country does not address the inequalities already inherent in their schooling as reflected in Figure 10.4, increasing demanding OTL in their standards would most likely increase the magnitude of the indirect effect, thus increasing the total SES effect on performance. This in many ways is the ignored or at least somewhat hidden irony of school reform – pursuing

What Has Been Learned about the Role of Schooling

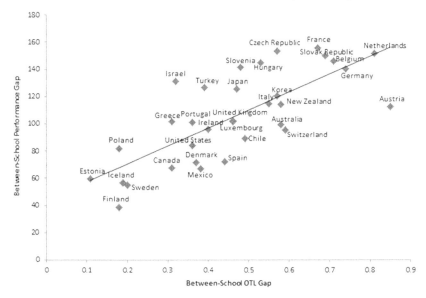

Figure 10.5 Relationship across countries of size of OTL gap and performance gap at the between-school level

excellence can at the same time result in greater inequalities. If the more demanding curriculum standards defining content coverage (OTL) are not available to all students but are introduced differentially to subgroups of students, then the effect on inequality grows larger. Metaphorically, the rich get richer and the poor get poorer.

A parallel argument can be made for looking at differences in both excellence and equality across countries. Those countries whose curriculum standards are more rigorous, focused, and coherent – the characteristics found to be related to higher performance in mathematics and science in TIMSS-95 (see Chapter 7) – would ironically seem to have the greatest potential for higher levels of inequality.

Given the results displayed in Figure 10.4, where on average across the OECD countries about one-third of the total SES inequality is related to schooling (OTL) and assuming the OECD average size of the SES to OTL estimated effect as true for all countries, the countries with the more coherent and rigorous curriculum resulting in higher performance would correspondingly have greater inequality. This would be true unless those countries had a smaller value for the size of the relationship between SES and OTL. Lower inequality can occur in the presence of excellence in performance but requires supporting policies and practices reducing the relationship between SES and OTL.

IX. *The quest for both excellence and equality in performance is attainable, although few countries have achieved it.*

Most countries have as an integral part of their national curricular standards a statement indicating that there are two major goals of education. One is the obvious goal of focusing on educational excellence as indicated by average student performance. This goal is what fuels the desire of nations to engage in international assessments such as TIMSS-95, TRENDS, and PISA. In this context, excellence is measured by a country's rank. But excellence is only one of the goals found in such policy documents. The other one is educational equality, which in more recent years has achieved increasing emphasis due to countries' concerns about the increasing income and wealth gaps found in their societies. They see schooling as the means to lessen such gaps and to give all children a chance at a better life no matter their current background.

Over thirty years ago, Oakes (1985) made the argument that these two goals are in competition with each other, producing a trade-off situation. More recently, the OECD using PISA-2012 data has focused on this issue, identifying countries as both high performers on the assessments as well as achieving high levels of equality (OECD, 2013b). That report identified nine OECD countries as meeting both criteria.

Using the analyses done in support of the results presented in Figure 10.4, a different approach to the identification of countries having both excellence and equality was taken (Schmidt & Burroughs, 2016). Seven measures of inequality were used. These included:

- Total SES effect (combining direct effect of SES on PISA scores and indirect effect of SES on math scores through the effect of SES on OTL);
- Direct SES effect (the coefficient of SES to PISA mathematics literacy, controlling for OTL);
- Indirect SES effect on PISA scores through the SES-OTL relationship;
- % Indirect (the proportion of total SES effect that was due to the indirect effect);
- SES-OTL (the estimated regression coefficient between SES and OTL);
- PerfGap (the difference in mean PISA scores between the top SES and bottom SES quartiles, with quartiles defined by the distribution of SES in each country);
- OTLGap (the difference in mean OTL between the top SES and bottom SES quartiles, defined as above) (Schmidt & Burroughs, 2016, p. 106).

Table 10.3 *Eleven countries with higher PISA mathematics scores and lower inequality*

	Total SES	Direct SES	Indirect SES	% Indirect	SES-OTL	PerfGap	OTLGap	% Variance
Australia		x						x
Austria		x						
Canada	x	x	x	x	x	x	x	x
Denmark	x		x				x	
Finland	x	x	x	x	x	x	x	x
Germany		x						
Ireland	x	x					x	
Japan		x	x	x	x	x	x	x
Korea		x					x	x
Netherlands	x	x				x		x
Switzerland			x			x		x

Note. Adapted from Table 2. Copyright 2016 by Georgetown University Press. From *Georgetown Journal of International Affairs: Winter/Spring 2016, Volume 17, No. 1*, Mike Fox and Angela Ribaudo, editors, p. 108. Reprinted with permission. www.press.georgetown.edu

The analyses were done with a focus on identifying countries for the United States to model, and as a result only twenty-seven of the thirty-three participating OECD countries were included. Of these only eleven had an above-average PISA-2012 mathematics performance and above-average equality on any of the seven equality measures (see Table 10.3) (Schmidt & Burroughs, 2016).

Three countries stood out in terms of equality, as they were found to be lower-than-average on all or almost all of the eight measures of inequality. Those countries were Canada, Finland, and Japan – all of which were also found on Schleicher's (2014) OECD list. The other eight countries were all above-average on the excellence dimension and were also above average on at least one of the equality measures.

X. *The eighteen-country study of teacher preparation for future lower-secondary mathematics teachers, Teacher Education and Development Study in Mathematics (TEDS-M), found that the content coverage in the mathematics courses taken by a majority of the students in the best programs – defined as those having the highest average level of performance on a test administered before graduating from their teacher preparation program – included significantly more advanced mathematics, including calculus, analysis, differential equations, and abstract algebra, than those in the remaining programs.*

Although TEDS-M is not a performance assessment of students and was not discussed in the previous chapters, we included this finding because of its relevance to student performance (Cogan, Schmidt, & Houang, 2014; Schmidt, Burroughs, Cogan, & Houang, 2017; Schmidt & Cogan, 2014; Schmidt, Cogan, & Houang, 2014, 2011; Tatto et al., 2012). IEA studies and PISA-2012 included measures of the curriculum (OTL) but the other major variable in the schooling equation is teacher quality, which can be operationalized as knowledge of mathematics and knowledge of how to teach mathematics.

Teacher knowledge is the key variable missing in all current international assessments. IEA studies have always included a teacher questionnaire that asks questions about teacher background, professional experience, attitudes, and instructional practices. PISA has not had a teacher questionnaire but does have a separate study of teachers (TALIS) asking similar questions. There is consideration within OECD to integrate TALIS and PISA in the future. Neither, however, measures teacher knowledge directly nor their knowledge about how to teach.

The IEA TEDS-M study addressed this issue by assessing the knowledge of future teachers during their final semester of teacher preparation. It is included here among the nine major findings of international assessments as a window into the important role of teachers in schooling. The findings of the strong relationship of OTL to performance for students is also true for teachers. Linking TEDS-M results with TIMSS-95 results shows that countries with stronger curricula (OTL) also have stronger curricula for the preparation of mathematics teachers (Schmidt, Houang, & Cogan, 2011).

The OECD TALIS study also looked at various aspects of teachers, ranging from their training, the support they receive from school leaders, and their practices in the classroom. It is the study components about classroom practices that explore the role of teachers in student learning, delving into the metaphorical black box to understand how the teaching–learning relationship between teacher and student plays out in daily practices. In the TALIS report (OECD, 2014a), the chapter dedicated to teachers' practices in the classroom begins by acknowledging the relationship, stating "[q]uality instruction encompasses the use of different teaching practices, and the teaching practices deployed by teachers can play a role in student learning and motivation to learn" (pp. 150). This introductory sentence is followed by an overview of the research surrounding the relationship, including a brief discussion of the importance of a positive classroom climate in which student motivation and learning is cultivated by teacher efforts to develop a safe and supportive classroom that limits disruptive behavior, thus allowing for more teaching time.

While the TALIS report acknowledges the relationship between teaching and learning, the nature of the TALIS study is "to examine teaching practices that teachers report using in the classroom and how these practices relate to the beliefs that teachers hold and the environments in which teachers work" (pp.150). Thus, since the data is teacher-reported practices and beliefs and the study is not designed to understand the student side of learning – achievement, motivation to learn, etc. – the most TALIS can do in exploring the teaching–learning relationship is develop a framework of the relationship from a teacher-centered perspective.

Teachers were asked to report on their use of eight different classroom practices, but the TALIS report focuses on three practices that are considered "active teaching practices" – or strategies in which students are central actors in the learning process – since these strategies may be more effective in engendering student learning depending upon how they are implemented by the teacher. These three practices are: (a) students work on projects requiring at least one week to complete; (b) students use information and communication technology (ICT) for class work or projects; and (c) students work in small groups to come up with a solution to a task or problem.

The variation in the cumulative percentage of all three of these practices is explored in the TALIS report by looking at factors at the country, school, and teacher levels, yet the variance was found to be mainly due to individual teacher factors rather than school or country factors. Interestingly, while the TALIS report points to the teacher as an individual as the main reason that different practices are used, the report also found that school and country factors play a larger role in the use of practices requiring substantial resources (such as the use of ICT). However, the role of school and country factors was still not as substantial as the role of the individual teacher in accounting for the variance.

The TALIS report also looked at teacher use of student evaluation, teacher beliefs about their role in student learning, cooperative and collaborative practices between teachers, and the classroom environment. Like the previous variables, the variance for the majority of these variables was found at the teacher-level; the exception lies in cooperative and collaborative practices between teachers, which have a greater country-level variance than any other of the teacher practice variables. While the TALIS report says this may be a cultural factor, it also says that collaboration is best promoted by school leaders, so it is not only related to teacher practices but also leadership practices.

The overarching recommendations of TALIS are that teachers need to be supported in developing active teaching practices, collaborative behavior needs to be promoted by school leaders, and teachers need more professional development to develop better classroom management skills in order to have more time devoted to teaching and learning.

These represent the top ten findings related to the role of schooling in student learning. We conclude this chapter with some related findings in support of the ten points. They come from three specific countries – Russia, Germany, and the United States.

What We Can Learn from Russia, Germany, and the United States

A Unique Opportunity: The Russian Longitudinal Study

Backing up in time to the years 2011 and 2012 provides us with a rare opportunity to explore the issue of the relationship of OTL and SES to performance with a unique set of data in which the same students took both the TRENDS-2011 and the PISA-2012 assessments.

Russia participated in TRENDS-2011 at the eighth grade, drawing a sample of 4,893 students in 231 classrooms in 210 schools in 50 regions. In the following year Russia also participated in PISA-2012 by drawing a traditional sample of fifteen-year-olds. In addition, they also administered the PISA assessment to the students who had participated in TRENDS-2011, successfully following up on 90 percent of the original sample (including 4,399 students in 229 classrooms in 208 schools). This created a true longitudinal data set of students that took the TRENDS assessment and then one year later took the PISA literacy assessment.

What makes these data even more relevant to exploring the substantive issue of schooling is that when these students took the TRENDS-2011 assessment they were in eighth grade (the modal age being fourteen) and the following year when they took the PISA-2012 they were ninth graders who were fifteen years of age. The continuity of these students moving from eighth grade to ninth grade thus having one more year of schooling made analytical possibilities much more interesting, especially as it relates to OTL and its relationship to performance. The overall relationship between the two tests at the school level is illustrated in Figure 10.6.

Interestingly from a methodological point of view, the regular PISA-2012 sample for Russia also provides a cohort longitudinal data set. Making the assumption that the cohort of eighth graders in 2011 was demographically

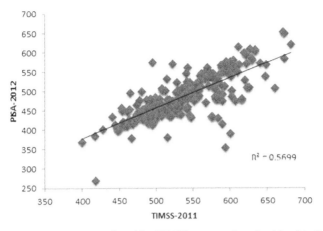

Figure 10.6   PISA-2012 predicted by TIMSS-2011 at the school level in Russia

and in other ways similar to the cohort of ninth graders in 2012 with the exception of one more year of schooling, the data allow a comparison at the country level that is longitudinal (eighth grade to ninth grade) by treating the eighth-grade aggregate score at the country level as a pre-test for the ninth graders. This can then be compared to the country-level results from the true longitudinal study. In this section we examine the issues of the relationship of OTL and SES to performance with the Russian longitudinal data.

A recent paper by Carnoy, Khavenson, Loyalka, Schmidt, and Zakharov (2016) used the Russian true longitudinal data to study the relationship of OTL and teacher quality to PISA performance. The conclusion reached was that the positive relationship of teacher quality and opportunity to learn to performance estimated through longitudinal data was more modest than from cross-sectional data. The study did find a statistically significant relationship ($p < .01$) for the PISA OTL measure of formal mathematics controlling for SES, school type, and several variables describing the teacher in relationship to PISA performance with an estimated standardized regression coefficient of .15. This analysis essentially replicates findings for the PISA cross-sectional analysis in terms of a significant relationship of OTL to performance.

The paper goes on to present the results of the same analysis but also controlling for previous achievement using the TRENDS-2011 eighth-grade scores (see Table 10.4). That analysis also found a statistically significant relationship of formal mathematics OTL to PISA performance ($p < .01$). The size of the standardized coefficient, although still significant,

Table 10.4 Estimated student achievement, PISA-2012, including TRENDS-2011 mathematics score[2]

	Model 1	Model 2	Model 3	Model 4	Model 5	Model 6
TIMSS Math Score 2011	0.53***	0.53***	0.53***	0.52***	0.53***	0.52***
Female	-0.08***	-0.08***	-0.08***	-0.06*	-0.06*	-0.06*
Class Average BIH (% > sample median)	0.09***	0.08***	0.08***	0.08***	0.08***	0.08***
School Type: Gymnasium	0.18***	0.16***	0.15*	0.14*	0.14*	0.13*
School Type: Lyceum	0.17*	0.19*	0.19*	0.21*	0.21*	0.19*
School Type: Educational Center	-0.21	-0.18	-0.20	-0.31**	-0.30**	-0.27*
Teacher Preservice Math Education		-0.14**	-0.15**	-0.15*	-0.15*	-0.15**
Teacher Preservice No Math Education		-0.17	-0.17	-0.21*	-0.20	-0.20
Teacher Years Teaching Math		0.01	0.00	0.00	0.00	0.00
Years Teaching Math Squared		-0.00	-0.00	-0.00	-0.00	-0.00
Teacher Highest Category			0.00	0.00	-0.02	-0.03
Teacher Second Highest Category			0.04	0.04	0.03	0.02
Teacher Lowest Category			-0.17	-0.22	-0.21	-0.23
Workload Classes				0.00	0.00	0.00
Workload Out of Classes				0.00	0.00	0.00
Workload Administration				0.00	0.00	-0.00
Exposure Applied Math (z-score)				-0.07***		
Exposure Word Problems (z-score)					0.01	
Exposure Formal Math (z-score)						0.09***
Constant	1.60**	1.69**	1.68**	1.17	1.22	1.11
Control for Student FAR	Yes	Yes	Yes	Yes	Yes	Yes
Observations	4,389	4,389	4,389	2,908	2,901	2,920
Adjusted R^2	0.437	0.440	0.442	0.437	0.431	0.441

Source. Russia TIMSS-PISA sample, 2011–2012.

Note. Reference variables: 0–10 books in the home; mother's education = high school complete; teacher preservice education = degree in mathematics; teacher third highest category; school type - regular secondary school. Standard errors of coefficient estimate available on request. HS = high school; BIH = books in home; TIMSS = Trends in International Mathematics and Science Survey; PISA = Program for International Student Assessment; FAR = family academic resources.

*p < .10; **p < .05; ***p < .01.

Note. Adapted from "Revisiting the Relationship Between International Assessment Outcomes and Educational Production: Evidence From a Longitudinal PISA-TIMSS Sample" by M. Carnoy, T. Khavenson, P. Loyalka, W. H. Schmidt, & A. Zakharov, 2016, *American Educational Research Journal*, 53, p.1076. Copyright @ 2016. Reprinted by Permission of SAGE Publications, Inc.

[2] Exposure Formal Math in the table refers to the formal mathematics OTL measure in Figure 1.3,4b of the PISA 2012 Volume I report, indicating the amount of coverage related to more advanced mathematics (OECD, 2012).

was smaller (.09), which led the authors to conclude: "This study shows that the cross-sectional estimates behind such recommendations may be biased ... Our results suggest that the positive roles of teacher quality and 'opportunity to learn' in improving student performance are much more modest than claimed in PISA documents" (Carnoy, Khavenson, Loyalka, Schmidt, & Zakharov, 2016, p.1).

However, the alternative explanation is that the inclusion of the eighth-grade TRENDS performance score as an additional control variable removes some of the cumulative OTL experienced by students up to and including that experienced in the previous year. Controlling for prior achievement narrows the interpretation of the estimated PISA OTL effect to the opportunities experienced in ninth grade. Note also that the OTL variables in this analysis were the only ones converted to z-scores, making comparisons of the regression estimates problematic.

Using TRENDS-2011 performance as a control variable in the analysis of PISA performance can also be modeled as a structural equation reflecting the conception that performance on the school-based curriculum-oriented assessment reflects not only a direct effect but is also part of the indirect effect of eighth-grade OTL on ninth-grade performance in Russia.

The model shown in Figure 10.7 hypothesizes that eighth-grade OTL has an indirect effect through TRENDS performance to ninth-grade performance on PISA.[3] The direct effect of eighth-grade OTL on eighth-grade performance, if significant, would indicate that the inclusion of TRENDS-2011 performance as a covariate in the model as done by Carnoy et al. (2016) would have the effect of reducing the magnitude of the PISA OTL effect on PISA performance, as the eighth-grade OTL effect on PISA performance is embedded in the ninth-grade PISA OTL effect on performance. Given the cumulative nature of literacy performance, the total OTL effect must include the exposure over all grades prior to and including ninth grade not just that ninth-grade exposure, which is what happens when previous OTL is at least partially removed by the inclusion of the TRENDS performance. We hypothesize that is what, at least in part, caused the reduction in the effect size of OTL in the Carnoy et al. (2016) paper.

The results of fitting the model provide support for the hypothesis. Eighth-grade OTL is statistically significantly related to eighth-grade TRENDS performance, implying the existence of an additional indirect

---

[3] The model is not fully specified as it is intended to refer to that part of the Carnoy et al. (2016) model related to OTL that was presented in Table 10.4.

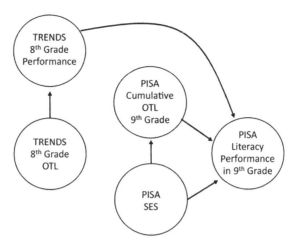

Figure 10.7 Model for Russia relating OTL and eighth-grade performance to ninth-grade performance

OTL effect on PISA performance in addition to that defined by the direct effect of PISA OTL to performance. Since PISA literacy is a cumulative measure, both sources of OTL should be included. This supports the notion that including eighth-grade performance as a control variable, as done in the Carnoy analyses presented in Table 10.4 to determine the size of the schooling effect (OTL-content coverage) on PISA-2012 performance in mathematics literacy, is too restrictive, since the PISA OTL measure is cumulative and the inclusion of eighth-grade performance is removing some of that cumulative OTL, at least that part reflected in the indirect relationship of eighth-grade OTL to PISA ninth-grade performance through the relationship of eighth-grade performance to ninth-grade performance.

We now turn to Germany and the United States as two countries that have a federal system of government, particularly with respect to the organization of their educational systems. Germany has sixteen Länder, while the United States has fifty states, and in both countries most educational policy is determined at that level. In the previous section we focused on what we can learn about the fundamentals of schooling from international educational assessments of mathematics, given the salient role that it has come to play in the technologically oriented economy and in the data-driven societies in which we live. In what follows, we examine how these two countries responded to PISA and IEA findings by introducing policy changes that transformed their systems in a significant way.

United States

For the United States, the results that in particular influenced policy changes came from the TIMSS-95 curriculum analysis.[4] The achievement results also played a role by pointing out how poorly US students did, especially at the eighth and twelfth grades. Catching the attention of the nation and key policy makers was made possible by the widespread coverage of both of these aspects of the TIMSS-95 study by much of the nation's press, including major television outlets such as CNN, NBC, and ABC.

The White House, under the direction of President Clinton, and the two houses of congress asked for briefings and the White House, using the presidential platform, called for action to reform mathematics and science education as well as to introduce a policy initiative related to a voluntary national examination that failed to become law in the Congress.

Prior to TIMSS-95 there had been a growing movement for the strengthening of state standards. The results from TIMSS-95 emboldened that movement and encouraged not only strengthening the standards but also making them national (not federal) in scope (Schmidt, McKnight, Cogan, Jakwerth, & Houang, 1999; Schmidt, McKnight, & Raizen, 1997). TIMSS-95 results describing the key features of the standards of the top-achieving countries – coherence, focus, and rigor as discussed in Chapter 7 – were released in 2001 (Schmidt, Wang, & McKnight, 2005). Based on these results, states began to change their state standards to reflect these principles. Notable were Minnesota and Massachusetts in their efforts to strengthen their standards. Minnesota's efforts in this regard resulted in their performance in mathematics on the 2007 TRENDS study showing a much larger gain than was the gain for the United States as a whole (SciMath^{MN}, 2008). Other states responded as well, and there were several efforts on the part of the nonprofit Achieve organization to develop common examinations, so as to encourage higher standards.

Two other policy documents occurred during this same time period that acknowledged the TIMSS-95 research results. The National Council of Teachers of Mathematics (NCTM) put out a publication for teachers defining the key focal points that should be emphasized in the mathematics K–8 curriculum so as to realize the TIMSS-95 principles of focus and coherence. In 2008 President George W. Bush organized a National Mathematics Advisory Panel to suggest policies for improving the "performance in mathematics

---

[4] What follows is a summary of the Schmidt and Burrough's 2016 paper.

among American students" (National Mathematics Advisory Panel, 2008, xiii). The panel responded by stating in the executive summary that:

> A focused, coherent progression of mathematics learning, with an emphasis on proficiency with key topics, should become the norm in elementary and middle school mathematics curricula. Any approach that continually revisits topics year after year without closure is to be avoided. (NMAP, 2008, p. xvi)

The panel further stated:

> By the term focused, the Panel means that curriculum must include (and engage with adequate depth) the most important topics underlying success in school algebra. By the term coherent, the Panel means that the curriculum is marked by effective, logical progressions from earlier, less sophisticated topics into later, more sophisticated ones. (NMAP, 2008, p. xvii)

Laying fallow over this period of time were the seeds laid out in *A Splintered Vision* (1997) that for the United States to really reform US mathematics education it was necessary to have a national curriculum – one that was directed by the states themselves and not the federal government. The germination of those seeds occurred around 2009 when the National Governors Association, together with the Council of Chief State School Officers, with the support of Achieve, came together to create what were called the Common Core State Standards for Mathematics (CCSSM). The name has been changed to College and Career Ready Standards to alleviate political issues, but over forty-five states, with various small modifications, have adopted these standards in substance if not in name. The two groups driving this effort are both state-level organizations representing the two offices most responsible for education in the United States from each state – the governor and the state superintendent of education.

This was one of the most profound policy changes in American educational history. It ran afoul of the time-standing tradition of having standards that are, at most, state defined and, for many, district or even school defined. The new standards were not only different in being national but were explicitly based on the TIMSS-95 research findings that they be focused, coherent, and rigorous. Specifically quoting from the US report, *A Splintered Vision*, the authors of the CCSSM in their introduction noted:

> For over a decade research studies of mathematics education in high-performing countries have pointed to the conclusion that the mathematics curriculum in the United States must become substantially more focused and coherent in order to improve mathematics achievement in this country.

> To deliver on the promise of common standards, the standards must address the problem of a curriculum that is a "mile wide and an inch deep" ... (National Governors Association Center for best Practices, & Council of Chief State School Officers, 2010, p. 3).

Assessing the coherence of a set of standards is more difficult than assessing the focus. Schmidt and Houang (2002) have said that content standards and curricula are coherent if they are "articulated over time as a sequence of topics and performances that are logical and reflect, where appropriate, the sequential or hierarchical nature of the disciplinary content from which the subject matter derives" (p. 9).

Although not with as dramatic a history or as much national attention and fanfare, the development of the new national science standards – Next Generation Science Standards – were also influenced by the TIMSS-95 curriculum analysis results. Again quoting from *A Spintered Vision*, the National Research Council Report (2012) stated:

> The framework is motivated in part by a growing national consensus around the need for greater coherence – that is, a sense of unity – in K-12 science education. Too often, standards are long lists of detailed and disconnected facts, reinforcing the criticism that science curricula in the United States tend to be "a mile wide and an inch deep" [1] (National Research Council, 2012, p. 10).

Germany

Similar to the United States, Germany was concerned about the performance of its students on the international assessment. "When the results became public in Germany, there was grave concern that its once highly regarded, even revered, education system was deteriorating. Policymakers and analysts also worried that there were serious flaws in the structure of the system itself, flaws that had potential negative consequences, both economically and socially" (Schmidt, Houang, and Shakrani, 2009, p. 11). In reaction to those results from TIMSS-95 and PISA, "a number of steps have been taken towards the development and monitoring of educational quality in Germany" (Pant, Tiffin-Richards, & Stanat, 2017, p. 51).

Also similar to the United States, as noted previously, is the structure of the educational system. Education is mostly under the purview of the sixteen Länder (states). As in the United States, the federal government has no constitutional authority over education. Germany has a conference of state education ministers named the Ständige Konferenz der Kultusminister der

Länder (KMK). They are charged with the task of coordinating education issues in addition to research and cultural affairs. In 2003 and 2004, the KMK introduced national standards and standardized testing procedures (Schmidt, Houang, and Shakrani, 2009). "This decision clearly marked the starting point for Germany to monitor and control the outcomes of its education system at the national level, a monumental change in a country that had operated under state control" (Schmidt, Houang, & Shakrani, 2009, p. 11).

The standards introduced were:

> ... for the primary level and for secondary level 1 detailing which competencies and skills students are expected to have developed by the time they reach certain points in their school career. In general, the key objective of introducing National Educational Standards was to shift attention to the learning outcomes of educational processes and to ensure greater comparability of educational requirements within the school system.
>
> At the primary level, the focus of these standards was on the core subjects of German and mathematics. At secondary level I, the focus was on German, mathematics, the first foreign languages (English, French), and for the science subjects biology, chemistry, and physics. In accordance with the long-term strategy of the Standing Conference (KMK 2015) on educational monitoring in Germany, the 16 federal states (Länder) also decided to conduct regular comparative studies using standardized tests on three levels (Pant, Tiffin-Richards, & Stanat, 2017, p. 51).

So, in many ways, Germany's story is very similar to that of the United States, where the results of the international assessments moved both countries into new policy terrain from where they had previously been – both moving in a more national way than they had previously. The authors, not being German, had to rely heavily on the published document cited earlier, and in fact we chose to quote the article at length so as to authenticate what was being said in the German authors' own words. There is also probably a richer context to these developments, which we are unable to provide as we did for the United States only because we lack such detail.

CHAPTER 11

Where Do We Go from Here?

In this book, we have examined the sixty-year historical development of international educational assessments focused on mathematics and science – looking at how they were designed, carried out, and analyzed; indicating by whom they were done; and how they have influenced educational policy and practice. From the beginning they were built upon the relatively simple idea of several university professors, who felt that such quantitative comparative data about schooling would lead to improving education. But at the same time they also recognized the potential pitfalls inherent in cross-country comparisons. All the seeds of what we know and do today in TIMSS and PISA were planted back in the early 1960s. What has evolved is much more complex and methodologically sophisticated, supported by technological advances, especially in computing, which occurred over that same span of time.

Having spent a considerable part of our professional careers working with international assessments leads us in this chapter to share our vision of what we recommend and would like to see in the future generations of such studies. Instead of looking back over the chapters and summarizing what has been, we choose in our final chapter to look forward to what is yet to be. What follows are the insights and ruminations of three of the authors – Schmidt, Houang, and Cogan – as our younger co-author shares no blame in what follows. And what follows is not intended as a critique of what has been done but rather a forward look as to what might be.

The exercise of looking back into the history of such studies to write the first chapters of this book made us realize that much of our self-proclaimed "brilliance" in what we have done more recently was simply to provide further conceptualization and methodology for the seeds sown back in the 1960s. In the same vein, we hope that what we have learned may similarly inform the future.

Very large sums of money and human effort go into the OECD and IEA studies. Each of the studies is done professionally and with meticulous care. We learn something each cycle but the question remains: what more could we learn if these studies were organized and designed differently? When Einstein was confronted with a perplexing problem, he would conduct a thought experiment. We decided to follow suit and ask the question of what we might gain with a new approach to conducting international assessments of education.

We also have chosen not to take the obvious approach of ignoring the history and expertise related to PISA and TIMSS and invent a new single study (in this chapter we refer to TRENDS simply as TIMSS, as the distinction with TIMSS-95 is no longer important). Instead, we choose to keep the heart of the two studies and address how they might be better coordinated. We also propose how greater coordination of the two studies could provide more useful information for the participating countries' educational systems.

The motivation for the greater coordination of PISA and TIMSS studies rests in our minds on the relationship of mathematical literacy and formal school-related mathematics proficiency. The degree to which they are related speaks to the importance of greater coordination. If they are two disparate entities then the benefits of coordination are small, as they tell two different stories, but if they tell two related stories then coordination could add value. We examine how the two have come together in today's world:

> For most countries, mathematical literacy is an expected outcome of schooling. This has been true for a long time. Mathematical literacy initially encompassed basic arithmetic skills such as adding, subtracting, multiplying, and dividing whole numbers, decimals, and fractions; computing percentages; and computing the area and volume of simple geometric shapes. More recently, the digitization of many aspects of life, the ubiquity of data for making personal decisions involving health and investments – as well as major social, economic, environmental, and policy decisions – have all reshaped what it means to be mathematically literate and to be prepared to be a thoughtful, engaged, and reflective citizen.
>
> Manufacturing is particularly illustrative of this dramatic change. The skills required of people in the new high-tech manufacturing sector are not what they were twenty years ago. Countries with advanced economies still require a manufacturing workforce, but those individuals are not doing the assembling. They are maintaining complex websites, tracking inventory using sophisticated software, coding to fix problems, and operating computerized assembly machines and robots – all of which require a level of proficiency in mathematics that was not needed in yesterday's manufacturing plants. The old adage that all that is needed to work in manufacturing is

a strong back is no longer true. A strong *background* in technology and mathematics is now necessary ...

In other words, there is a new "basic" for basic skills, demanding a reconsideration of what it means to be "literate" in mathematics, including the role of twenty-first century competencies as they relate specifically to mathematics (Schmidt, 2017, pp. 32–33).

These concerns have led to a broadened perspective on mathematical literacy that emphasizes: (1) the core of mathematics – mathematical reasoning informed by a small set of fundamental mathematics and statistics concepts that transcend and undergird the particulars of mathematics topics, such as the concepts of functional relationships in mathematics and variance in statistics; (2) more contemporary aspects of mathematics, such as computer simulations and discrete modeling, that were not emphasized in the last century, and (3) the twenty-first-century skills or competencies recognized by business and industry as the new basics, specifically those that are related to mathematics and statistics and their canons of inquiry (National Research Council, 2012; PISA 2021 Mathematics Advisory Board, 2017; Reimers & Chung, 2016). In this way, literacy and formal school mathematics can be viewed as two lenses on mathematics proficiencies and how they relate to twenty-first-century skills.

Toward Greater Coordination

Both TIMSS and PISA have a role to play as they examine different yet related aspects of student mathematics performance – one more focused on what was learned about the abstract system of mathematics and the other on how the knowledge, skills and reasoning related to that formal system can be applied to the real world.

If we look at the number of countries that participated in both TIMSS and PISA, one might well conclude that both are not needed. In 2015, for example, when countries had the opportunity to participate in both assessment programs, only about one-third of all those that participated in at least one of the two participated in both (see Table 3.1). For these twenty-nine countries, the regression analysis relating TIMSS performance to PISA performance revealed a strong relationship. At the country level TIMSS performance accounted for nearly 90 percent of the variance in PISA performance (see Figure 11.1).

Moreover, in the original 1995 Third International Mathematics and Science Study, forty-four countries participated at eighth grade, including most of the European countries. Many (thirty-one) of these same countries

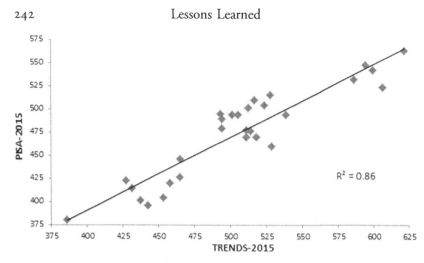

Figure 11.1   Predicting PISA-2015 from TRENDS-2015

also participated in the first mathematics focused PISA in 2003. However, of these thirty-one countries only seventeen (39 percent) also participated in TRENDS-2003 in that same year.

In spite of such lack of participation in both studies when the opportunity was presented and the high correlation between the two country-level scores, other types of data suggest the importance of both. The 2012 PISA study showed the strikingly strong relationship of opportunity to learn formal school mathematics (OTL) and PISA performance on the mathematics literacy assessment. This relationship was statistically significant for all participating countries (see Chapters 6 and 10).

The OTL measure indicated the degree of emphasis students reported they had experienced cumulatively over their schooling on formal mathematics content areas. The existence of such a strong relationship suggests that formal school mathematics is likely foundational to developing mathematical literacy. This is at least a hypothesis worth exploring. From IEA studies in general and particularly from TIMSS-95 we know there is also a strong relationship of formal school OTL to the TIMSS school curriculum-focused assessments.

Perhaps the most convincing data that both assessments are important for understanding mathematics learning and are not redundant comes from the Russian Federation (Russia). As described in Chapter 10, Russia followed up the eighth-grade sample of students who were in TRENDS-2011 and had them also take the PISA-2012 literacy assessment. As reported by Carnoy, Khavenson, Loyalka, Schmidt, and

Zakharov (2016) and the additional analyses reported in Chapter 10, the regression analysis relating TIMSS performance to PISA performance accounted for only about half of the variance in PISA performance. This estimated relationship leaves a substantial amount of variance in PISA performance to other factors, suggesting that the two tests characterize different yet related aspects of student mathematics learning. In addition, Russia is one of several countries, including the United States, that perform at different levels on the two assessments in cross-country comparisons, providing further evidence that the two programs are tapping into different aspects of mathematics competencies.

Combining TIMSS and PISA results creates a more complete picture of the relationship between formal school OTL and mathematics performance. The relationship between these two may seem to be common sense with respect to the more school curriculum-based TIMSS assessment but is really quite remarkable with respect to the mathematical literacy focus of PISA. Of particular interest is the role that formal school OTL may have in the development and practice of mathematical reasoning portrayed in the PISA model of problem solving (see Figure 4.1), in which mathematics reasoning transforms the messy, poorly defined real-world problem involving quantities into a well-defined mathematics problem that may be solved mathematically. TIMSS and PISA both conceptually and empirically provide insight into the contribution of formal school OTL to the development of different aspects of student mathematics performance. This is something that deserves further research and exploration. Coordination of the two studies could be of great value in examining this issue.

Providing a Window on K–12 Schooling

Participating in both PISA and TIMSS provides a country-level window across the major segments of K–12 schooling – primary, lower-secondary (middle school) and upper-secondary (high school).

PISA focuses on age as the basis for the definition of the population to be sampled, while TIMSS chooses the grade in which the largest number of students of a particular age are to be found. For TIMSS a single grade – usually eighth grade (also fourth as a separate population) for most countries – becomes the population. In contrast, the PISA student sample typically extends across three or more grades in order to represent the fifteen-year-old population (see Chapter 3). This provides another justification for countries that can afford to do both TIMSS and PISA, for when the results are combined at the country level the data characterize three major segments of K–12 schooling – fourth grade, representing primary school, eighth grade, representing lower-secondary school, and fifteen-year-olds, representing upper-secondary school (ranging from some combination of ninth, tenth, eleventh, and twelfth grades – see Table 3.1).

Originally, focusing on fifteen-year-olds made sense as it was the age defining the end of compulsory education for many countries. A side question arises: is this still relevant enough in today's world both economically and socially, especially given that more countries are moving to the expectation of twelve years of compulsory schooling for all students? In this context it might also make sense, even if the population is defined as fifteen-year-olds, for PISA to move to a single grade or perhaps the two grades having the most fifteen-year-olds. This seems particularly relevant given the result that schooling (OTL) was related to PISA performance. This would make school-related analyses more interpretable in relation to school-based content coverage. In most countries fifteen-year-olds can span three or more grade levels, producing within-country results with large OTL differences.

Toward Country-Level Cohort-Longitudinal Studies

Coordinate the timing and sequencing of PISA and TIMSS to take advantage of possible cohort longitudinal studies at the country level across the three main segments of K–12 schooling.

For those countries participating in both PISA and TIMSS, results are available for student performance at three segments of the K–12 system. For the United States, for example, results would characterize performance on school-based formal mathematics at grades 4 and 8, and on mathematical literacy at grade 10 (the modal grade) as well as ninth and eleventh grades, although with much smaller samples. The fact that the formal schooling assessment results are available in the earlier grades is consistent with the results presented earlier regarding cumulative OTL and its relationship to mathematical literacy. The problem in using such results for cohort-type analyses is the alignment of the timing of when these studies occur. To make such cohort analyses more feasible requires a level of coordination between the two studies.

In making suggestions about the possible coordination between the two studies, we recognize the reality that some of the suggestions might not be

Where Do We Go from Here?

Figure 11.2 TIMSS and PISA schedule timeline for potential cohort longitudinal studies

realistic for historical or political reasons. One example is PISA's history and its defining signature around fifteen-years-olds no matter at which grade they are to be found. Note that this parallels the definition FIMS adopted for Population 1b (see Chapter 1). We respect those realities and acknowledge their relevance but still take the liberty of making such a bold suggestion as to what in our minds would produce more coherent results with respect to the role of schooling. What Russia did (Chapter 10) is a good example of the advantages that can be gained by having TIMSS (eighth grade) one or two years before PISA; ninth grade in the case of Russia or tenth grade in the case of two-thirds of the typical participating PISA countries.

The difficulty with this suggestion is that the most desirable aspect, the possibility of having concomitant or contiguous results, would only rarely occur given the difference between the PISA and TIMSS repetition cycles – every three versus four years. The obvious solution would be to have both studies on the same three- or four-year cycle and sequenced with a two-year gap between the two studies, with TIMSS occurring first because of the measurement of foundational mathematics knowledge that is related to the ability of students to apply it on a literacy assessment.

A third approach to this type of organization is to lay out a schedule where the aim is to assure that the TIMSS eighth-grade assessment occurs two years before PISA and leave PISA on its three-year cycle and TIMSS on its four-year cycle with slight adjustments for both. Figure 11.2 provides a look at the schedule for the next twenty years starting with 2019, where TIMSS will test the eighth graders and in 2021 PISA will test the tenth graders – the modal grade in two-thirds of the countries. This results in a cohort longitudinal study at the country level as the same cohort would be tested in two separate years, representing in most countries an eighth to tenth grade picture of cohort gain. However, only two such cohort situations are created over the next twenty years – one in each of science and mathematics. To create more such opportunities would require schedule adjustments.

What would be the advantage if such changes were made? It would produce more country-level cohort-longitudinal studies with the same cohort of students being tested when they were in eighth grade and then two years later when they are fifteen years old and, for most countries, in tenth grade. Longitudinal studies are very difficult to carry out as well as expensive but, as Husén noted, they are necessary if the goal is to identify country-specific factors related to performance. Having such continuity in performance and OTL for the cohort would provide valuable policy-relevant analyses. In the next section we include a set of observations that we believe would enhance the study of schooling worldwide.

Some General Observations

Sampling Issues

Sampling the student population should be designed so as to reflect the hierarchical and clustered structure of schooling, ensuring that all the most important levels are represented.

Both studies adhere very closely to standard survey sampling procedures, which provide unbiased estimates of the country means. The sample design, however, also impacts the kind of analyses that can be done with the data beyond estimating means and variances for countries as a whole. In accord with the founding father's concept of comparative education, we maintain that the purpose of an international assessment in education is to provide quantitative data that permit careful cross-country comparisons that contribute to our understanding of educational practices and policies that are related to student learning and ultimately their performance on an assessment, either one with a focus on literacy or one based on the school curriculum.

The sampling frames of both studies have difficulties in this respect. Consequently, variance estimates at different levels of the educational structure are confounded. It is important to be able to obtain unbiased estimates of the relevant variance components, in order to understand the implications of the findings and to model appropriately any relational analyses within any single country. This demands that the sampling include at least two if not more units at each level in the sampling design (students, classrooms, tracks, schools or school-types, and regions if desired).

For the IEA studies, given that the second stage sampling unit is the classroom so as to study variation in OTL, it would be ideal to select all classrooms at the chosen grade level within each sampled school and then to include all students within each classroom. Although such a procedure might be viewed as too costly and disruptive to the schools, it would permit unbiased estimates for the school variance as distinguished from classroom variance and both of these from the student-level variation. With the move to computerized testing, the costs of extra classrooms and grades might well be less than in previous studies.

As each source of variance has different policy and instructional implications, it is critical in a study of schooling to be able to obtain unbiased estimates at least for three levels, i.e. student, classroom, and school. PISA sampling of students confounds classrooms with students; TIMSS typically confounds classrooms with schools. In addition, if any type of tracking (see Chapter 9) is present then variance estimates for classrooms and tracks will be confounded as well in the TIMSS sampling design while variance estimates for tracking would be confounded with those for the classrooms and students in the PISA sampling design. As classrooms are, by definition, usually confounded with teachers and learning takes place in classrooms with the interaction of teachers and students around content, it is crucial to have unbiased estimates associated with classrooms/teachers. These considerations are important for analyses across countries but are absolutely critical if any one country is to gain insight about their own system.

For the fifteen-year-old population in PISA, sampling classrooms is quite difficult as secondary students of the same age in most countries are spread across three to four grades (see Table 3.1). This would also be true if a single grade or two grades were chosen for fifteen-year-olds. In that case it would be possible to draw at least a stratified random sample, where the strata are defined in terms of the program of study and tracks most frequently taken by students at that grade level in that school.

The sampling frame should also include the sampling of teachers of the chosen classrooms. In TIMSS, this is easily done, since the classroom is the second-stage sampling unit. For the fifteen-year-olds this is more difficult, yet what could be done is to obtain data from all the mathematics or science teachers of fifteen-year-olds or, if the proposed sampling scheme of only taking the modal grade were implemented, the teachers of the sampled classrooms could be included. To the degree to which PISA wants to relate formal schooling to literacy performance, the sampling of classroom and obtaining OTL data from teachers seems increasingly important. In fact, OECD is discussing the possibility of combining its study of teachers, TALIS, with PISA.

Longitudinal Studies of One Type or Another

Virtually all studies about schooling have the same goal: to demonstrate that schooling or various aspects of schooling are causally related to student learning. International assessments are not different. More studies, including longitudinal or cohort-longitudinal designs, are needed.

The first IEA study recognized the lure of causal interpretations from assessment results and cautioned against the misuse of these cohort-international studies to suggest causality. The Second International Study of Mathematics was the first international study to include a longitudinal sub-study in which eight countries participated. They focused on opportunity to learn and instructional practices and their relationship to what was learned at eighth grade controlling for a pre-measure of mathematics achievement.

In short, our recommendation is to do true longitudinal studies wherever possible. Having said that, we recognize that such studies are costly and they require an enormous amount of effort to follow up on the students over the course of a year. We recognize the desirability of such a design but realize its infeasibility. Hence, what we do recommend is that all TIMSS studies be done with a cohort-longitudinal design which can also be referred to as a synthetic-longitudinal design. The fact that this was done successfully in TIMSS-95 speaks to its feasibility and the advantage of being able to model gain and residual gain (Schmidt et al., 2001).

The costs are also not prohibitive, as it logistically requires testing additional classes with the same test at the same school. Computerized testing could greatly alter the cost structure. The extra burden resides primarily at the school level in terms of administering the tests. Staying with the traditional TIMSS focus, this would imply testing at the third and fourth grades and the seventh and eighth grades.

This, as discussed in Chapter 9, would enable a pre-posttest cohort-longitudinal design using the lower grade as the pre-test and the upper-grade for the post-test. The only assumption required by this cohort-longitudinal design is within-school demographic similarity across the two grades. Although not a true longitudinal study, the stronger the viability of the aggregate background homogeneity assumption, the closer this comes to approximating a longitudinal design at the classroom level.

The limitation is that the unit for the analyses is the classroom and not the individual student. However, since OTL is delivered at the classroom level by the teacher this is the appropriate unit of analysis. If tracking were present in the school then the sampling of the classrooms would need to be stratified with the tracks defining the strata.

If PISA were to go to the two adjacent grades in which were the majority of fifteen-year-olds as suggested previously, then the same strategy could be employed. Furthermore if PISA and TIMSS were to coordinate their testing schedules, the TIMSS testing at grade 8 could be used as the pre-test for PISA but only at the country level. The opportunity to follow up the TIMSS sample in PISA also is possible, as done by Russia in 2011 and 2012. The option with PISA as currently structured would be to draw a random sample of fifteen-year-olds from the same school, but the feasibility of this plan would depend upon what the modal grade would be and whether the two grades are in the same school. If not, the feasibility of this option would be prohibitive.

Opportunity to Learn

OTL must always be included in international educational assessment studies. Furthermore the measures should be defined at specific, not global, levels such as algebra, geometry, etc. – the more specific within the limits of response time, the more useful OTL becomes. Additionally, the measures should not merely indicate coverage versus no coverage, but rather an indication of the amount of time devoted to each topic. Finally, at least one measure of OTL at both the intended and implemented levels should be included. The intended level OTL should include all grades not just those being tested.

Without this type of measure, interpreting cross-country results is problematic and likely to yield misleading conclusions and implications, leading to erroneous educational policy. Potential causal explanations of cross-country differences in variables such as teacher experience, school structure, and attitude of the student toward mathematics are suspect and incomplete without controlling for OTL, as these factors are likely strongly correlated with OTL. Consequently, any estimated effects for these variables in the absence of control for OTL will be biased by an unknown amount and, therefore, problematic for informing appropriate policy.

Measures of OTL, in the same manner as measures of SES, must be included in all analytic considerations of student performance such as between countries; between schools within countries; between classrooms within schools; and between students within classrooms. This is not just because of conceptual interest in the estimated effects associated with schooling as measured by OTL and home/family background but because without proper control the relationship of other school-related measures to performance will almost undoubtedly yield biased results that lead to dubious suggestions for practice and policy. The probability of such biased effects increases with the size of the correlation between either OTL or SES with these other variables, i.e. the greater the correlation the larger the bias. OTL and SES conceptually represent the two most obvious fundamental influences on student learning and, as such, quantitative measures of these concepts are essential in *all* analyses related to student performance in the academic disciplines.

Having said this, the remaining question is how to measure OTL: who provides the OTL measure and in what metric? With regard to the metric issue, the more precise the measure of OTL, the richer the characterization of schooling that may be obtained, making possible the discovery of stronger relationships to performance. TIMSS-95 measured all three opportunity to learn levels, including intended, potentially implemented, and implemented (excluding the achieved curriculum from the discussion as it refers to measures of performance).

The results provided rich descriptions of the curriculum in participating countries, providing a set of variables useful not only to the interpretation of the assessment results but also providing a deep look at what was covered in the mathematics and science curricula of forty-some countries. One measure also provided an indication of emphasis and in what sequence the topics were covered across the grades for each country. Examining these data in the aggregate across the high-achieving countries yielded an international benchmark (A+ benchmark – see Chapter 7).

This type of extensive curriculum analysis we recommend be done on a rotating basis between mathematics and science, i.e. every five to ten years or perhaps every other TRENDS administration rather than at every cycle of TRENDS. We also recommend that both TRENDS and PISA always collect data on two specific measures of OTL. The first uses the standards documents (intended curriculum) as the basis to record the coverage of the topics across grades K–9 or K–12 depending on the population being studied. This is what was called Topic Trace Mapping in TIMSS-95, as described in Chapter 7.

The country representatives using the country standards documents and the most frequently used instructional materials (textbooks in most cases) records for each specified topic (derived from the study framework specifying exhaustively and exclusively the set of all mathematics topics or science topics for grades K–9, if not for grades K–12) the grades that the topic is covered, and the grade(s) at which it received its most focused coverage (see Chapter 7 for a description of what was developed in TIMSS-95). It is important that both the listed topics and the span of grades reflect all the grades at least up to the grade of the population being studied, e.g. at tenth grade for PISA and eighth grade for TIMSS. This is not equivalent to the traditional IEA test/item curriculum validation exercise as it is broader in scope and is not limited to the topics on the test.

This allows the development of the IGP, which, we believe, may have been one of the greatest contributions of the TIMSS-95 curriculum study to educational research in general. It provides a quantitative measure of schooling and represents a key component of the Carroll Model of School Learning – the content taught by teachers and responded to by students. This is the most functionally useful measure of the intended curriculum.

The second measure we recommend is the measure of how teachers allocated their instructional time over the school year to the coverage of each of a subset of the topics stipulated in the content framework. The chosen topics for the teacher questionnaire should be those that the Topic Trace Mapping showed on average across countries to be typically covered at the sampled grade level *plus or minus one grade*. Examples of those developed in TIMSS-95 for both mathematics and science are included in Appendix B. Again, this is essential to estimate the degree of emphasis on the focal grade-assessed topics.

As PISA moves to include teacher questionnaires, the same type of measures should be included. The Topic Trace Mapping should also become a part of PISA. However, if that were to be included, the grade range of the instrument should be expanded to include all the grades in which fifteen-year-olds are enrolled. In the short run before teacher questionnaires become a part of PISA and even after, a separate measure of OTL is recommended where the purpose is the same – topic coverage – but the informant is the student, as was done in PISA-2012.

At fifteen years of age not all students are taking the same courses. Unless PISA were to adopt our recommendation of focusing solely on the modal grade for fifteen-year-olds, any sampling of classrooms is essentially impossible. Without sampling entire classrooms, the only available informant on instructional content OTL is the student. The research and analytic purpose is the same but the nature and extent of the information gathered from students is different. The idea is students are given a set of topics or example items around a topic and asked how frequently in their schooling up to the present they have encountered/experienced/covered that topic. This was done for the first time in mathematics in PISA-2012 (example items are included in Appendix C). Given that this was the first time for this type of measurement, the methodology needs further development.

The other area of OTL measurement specifically related to PISA that we recommend records the school experiences students have in their classroom instruction and on the tests they have taken related to the application of mathematics (or science) to real-world types of problems. PISA is a literacy test and, as such, a measure of the degree to which students have had exposure to such types of experience is critical to understanding performance. TRENDS has also included applied problems in their assessment and, as a result, we recommend the inclusion of such applied OTL measures in both upcoming PISA and TRENDS studies. Here too, in PISA there were first attempts to obtain this type of OTL measurement, and so additional conceptual and methodological work is needed in this area.

Assessments: The Need for Content-Specific Sub-Scores

The more sub-test scores that are made available, the greater the opportunity to understand how schooling influences performance. This same principle applies to PISA although the sub-scores would be tied more to types of applications and to the different aspects of mathematical literacy.

Given the TRENDS focus on school-related learning, the assessments should be made curriculum sensitive both in the development of the items and in the creation of sub-scores. The total test score is not sufficient to shed light on the many important issues and policy questions related to schooling. The typical sub-scores of algebra, geometry, etc. are not much better at distinguishing performance as it relates to opportunity to learn. TRENDS should expand the number of sub-scores and make them more topic-specific at least at the country level using the percent correct metric.

The development of the test blueprints, the development of the items, the construction of the test forms, and the methodologically sophisticated IRT scaling are all state of the art techniques that are employed in both PISA and TRENDS. Our suggestion here is simple in conception, yet we recognize more difficult to implement. To estimate many sub-test scores depends on the test design and how many forms there are. As discussed in Chapter 4, in TIMSS-95 and in TRENDS, the typical total score and sub-scores in algebra, measurement, geometry, data, and number were estimated using IRT modeling. However, in TIMSS-95, twenty sub-test scores in mathematics (seventeen in science) were estimated in addition using the average percent correct at the country and classroom/school level (Schmidt, McKnight, Cogan, Jakwerth, & Houang, 1999).

The total test score, no matter how estimated, can be misleading, especially as to what it tells you about the origins of the variation in student performance. Sub-test scores are defined for smaller, more cohesive, and more homogeneous areas of mathematics and science and, as such, are more likely to reflect differences in content exposure (OTL). In contrast, total test scores measure a single underlying individual trait, typically defined as the student's general ability to learn mathematics, without reference to any particular area of mathematics. It is indicative of performance across such broad and differing areas as fractions, linear equations, quadratic functions, decimals, ratios, complex numbers, and rigid motions of geometric figures, to name a few. The type of latent trait underlying performance across such a broad spectrum of mathematics more typically reflects a student's natural ability in mathematics, SES, and/or his/her level of motivation in mathematics. As shown in Chapter 5, while country rankings were quite stable for the total score in mathematics and science, they varied wildly across the seventeen to twenty more specific sub-scores.

Maintain the Base while Introducing Variety

> *It would seem from the PISA and TIMSS results that major shifts in performance do not typically occur in any three- to four-year window. While maintaining the major trends in performance as well as the inclusion of the key variables of OTL and SES, PISA and TIMSS should include different specific topics or issues of interest in the fields of mathematics and science education for more intensive study in different cycles.*

PISA, given its focused testing of each of the three areas of reading, mathematics, and science every nine years, could choose to focus on a specific sub-area or topic each time mathematics and science are the focus. A specific area of mathematical literacy might get expanded coverage in the assessment and in the questionnaires, including expanded coverage of OTL. For example, while still retaining what is necessary to maintain the critical trends, deeper coverage might be given to statistical reasoning and how it works to support mathematical literacy. This would entail expanding items on the student (and teacher if PISA moves in that direction) questionnaire related to student OTL instructional experience, attitudes, and other related factors. The same principle could be applied to TIMSS in different areas of the curriculum where it is noted that students across countries have had major difficulties such as fractions at fourth grade.

Final Thoughts

TIMSS and PISA have evolved into a far more sophisticated endeavor to realize the goals of a small number of university researchers, who over sixty years ago had a simple but powerful and elegant idea about a quantitative comparison of schooling across countries and how those comparisons might inform nations about education reform. It began with a small number of countries (six), borrowed items from the University of Chicago, had small budgets, and made many assumptions. Although that pilot had many drawbacks and difficulties, it succeeded in laying the foundation of what today involves over one hundred countries, millions of students, large central staffs at OECD and IEA of psychometricians, sampling experts, researchers, computer programmers and other support staff, thousands of country-specific people engaged in collecting data and making the studies run, and budgets in the aggregate of over tens of millions of dollars.

Perhaps most amazing, quite unlike initial attempts at visibility in the public space, beginning with the 1995 Third International Mathematics and Science Study, the results of these studies now have worldwide visibility and enormous influence with governments, the education establishment, teachers, researchers, and policy analysts. As the lead author and one who was fortunate enough to have met and worked with some of those involved during the research and reporting cycles of SIMS, SISS, TIMSS-95, and PISA-2012, as well as to have been a student at the University of Chicago, where I was steeped in the tradition of where it all began many years before, I dedicate this book to honor the evolution of their simple yet far-sighted idea to a potentially ubiquitous and powerful influence on worldwide schooling but also to remind us all of their thoughtful concerns and cautions.

APPENDIX A

Third International Mathematics and Science Study 1995: Mathematics and Science Content Frameworks – Measuring Curricular Elements

TIMSS Mathematics Framework

Content
1.1 **Numbers**
    1.1.1 Whole Numbers
        1.1.1.1 Meaning
        1.1.1.2 Operations
        1.1.1.3 Properties of operations
    1.1.2 Fractions and decimals
        1.1.2.1 Common fractions
        1.1.2.2 Decimal fractions
        1.1.2.3 Relationships of common and decimal fractions
        1.1.2.4 Percentages
        1.1.2.5 Properties of common and decimal fractions
    1.1.3 Integer, rational, and real numbers
        1.1.3.1 Negative numbers, integers, and their properties
        1.1.3.2 Rational numbers and their properties
        1.1.3.3 Real numbers, their subsets, and their properties
    1.1.4 Other numbers and number concepts
        1.1.4.1 Binary arithmetic and/or other number bases
        1.1.4.2 Exponents, roots, and radicals
        1.1.4.3 Complex numbers and their properties
        1.1.4.4 Number theory
        1.1.4.5 Counting
    1.1.5 Estimation and number sense
        1.1.5.1 Estimating quantity and size
        1.1.5.2 Rounding and significant figures
        1.1.5.3 Estimating computations
        1.1.5.4 Exponents and orders of magnitude

1.2 **Measurement**
    1.2.1 Units
    1.2.2 Perimeter, area, and volume
    1.2.3 Estimation and errors

1.3 **Geometry: Position, visualization, and shape**
    1.3.1 Two-dimensional geometry: Coordinate geometry
    1.3.2 Two-dimensional geometry: Basics
    1.3.3 Two-dimensional geometry: Polygons and circles
    1.3.4 Three-dimensional geometry
    1.3.5 Vectors

1.4 **Geometry: Symmetry, congruence, and similarity**
    1.4.1 Transformations
    1.4.2 Congruence and similarity
    1.4.3 Constructions using straight-edge and compass

1.5 **Proportionality**
    1.5.1 Proportionality concepts
    1.5.2 Proportionality problems
    1.5.3 Slope and trigonometry
    1.5.4 Linear interpolation and extrapolation

1.6 **Functions, relations, and equations**
    1.6.1 Patterns, relations, and functions
    1.6.2 Equations and formulas

1.7 **Data representation, probability, and statistics**
    1.7.1 Data representation and analysis
    1.7.2 Uncertainty and probability

1.8 **Elementary analysis**
    1.8.1 Infinite processes
    1.8.2 Change

1.9 **Validation and structure**
    1.9.1 Validation and justification
    1.9.2 Structuring and abstracting

1.10 **Other content**
    1.10.1 Informatics

Performance Expectations
2.1 **Knowing**
    2.1.1 Representing
    2.1.2 Recognizing equivalents
    2.1.3 Recalling mathematical objects and properties

2.2 **Using routine procedures**
    2.2.1 Using equipment
    2.2.2 Performing routine procedures
    2.2.3 Using more complex procedures
2.3 **Investigating and problem solving**
    2.3.1 Formulating and clarifying problems and situations
    2.3.2 Developing strategy
    2.3.3 Solving
    2.3.4 Predicting
    2.3.5 Verifying
2.4 **Mathematical reasoning**
    2.4.1 Developing notation and vocabulary
    2.4.2 Developing algorithms
    2.4.3 Generalizing
    2.4.4 Conjecturing
    2.4.5 Justifying and proving
    2.4.6 Axiomatizing
2.5 **Communicating**
    2.5.1 Using vocabulary and notation
    2.5.2 Relating representations
    2.5.3 Describing/discussing
    2.5.4 Critiquing

Perspectives
3.1 **Attitudes toward science, mathematics, and technology**
3.2 **Careers involving science, mathematics and technology**
    3.2.1 Promoting careers in science, mathematics, and technology
    3.2.2 Promoting the importance of science, mathematics, and technology in nontechnical careers
3.3 **Participation in science and mathematics by underrepresented groups**
3.4 **Science, mathematics, and technology to increase interest**
3.5 **Scientific and mathematical habits of mind**

TIMSS Science Framework

Content
1.1 **Earth Sciences**
    1.1.1 **Earth Features**
        1.1.1.1 Composition

1.1.1.2 Landforms
1.1.1.3 Bodies of Water
1.1.1.4 Atmosphere
1.1.1.5 Rocks, Soil
1.1.1.6 Ice Forms
1.1.2 **Earth Processes**
1.1.2.1 Weather and Climate
1.1.2.2 Physical Cycles
1.1.2.3 Building and Breaking
1.1.2.4 Earth's History
1.1.3 **Earth in the Universe**
1.1.3.1 Earth in the Solar System
1.1.3.2 Planets in the Solar System
1.1.3.3 Beyond the Solar System
1.1.3.4 Evolution of the Universe
1.2 **Life Sciences**
1.2.1 **Diversity, Organization, Structure of Living Things**
1.2.1.1 Plants, Fungi
1.2.1.2 Animals
1.2.1.3 Other Organisms
1.2.1.4 Organs, Tissues
1.2.1.5 Cells
1.2.2 **Life Processes and Systems Enabling Life Functions**
1.2.2.1 Energy Handling
1.2.2.2 Sensing and Responding
1.2.2.3 Biochemical Processes in Cells
1.2.3 **Life Spirals, Genetic Continuity, Diversity**
1.2.3.1 Life Cycles
1.2.3.2 Reproduction
1.2.3.3 Variation and Inheritance
1.2.3.4 Evolution, Speciation, Diversity
1.2.3.5 Biochemistry of Genetics
1.2.4 **Interactions of Living Things**
1.2.4.1 Biomes and Ecosystems
1.2.4.2 Habitats and Niches
1.2.4.3 Interdependence of Life
1.2.4.4 Animal Behavior
1.2.5 **Human Biology and Health**
1.2.5.1 Nutrition
1.2.5.2 Disease

1.3 **Physical Sciences**
   1.3.1 **Matter**
      1.3.1.1 Classification of Matter
      1.3.1.2 Physical Properties
      1.3.1.3 Chemical Properties
   1.3.2 **Structure of Matter**
      1.3.2.1 Atoms, Ions, Molecules
      1.3.2.2 Macromolecules, Crystals
      1.3.2.3 Subatomic Particles
   1.3.3 **Energy and Physical Processes**
      1.3.3.1 Energy Types, Sources, Conversions
      1.3.3.2 Heat and Temperature
      1.3.3.3 Wave Phenomena
      1.3.3.4 Sound and Vibration
      1.3.3.5 Light
      1.3.3.6 Electricity
      1.3.3.7 Magnetism
   1.3.4 **Physical Transformations**
      1.3.4.1 Physical Changes
      1.3.4.2 Explanations of Physical Changes
      1.3.4.3 Kinetic Theory
      1.3.4.4 Quantum Theory and Fundamental Particles
   1.3.5 **Chemical Transformations**
      1.3.5.1 Chemical Changes
      1.3.5.2 Explanations of Chemical Changes
      1.3.5.3 Rate of Change and Equilibria
      1.3.5.4 Energy and Chemical Change
      1.3.5.5 Organic and Biochemical Changes
      1.3.5.6 Nuclear Chemistry
      1.3.5.7 Electrochemistry
   1.3.6 **Forces and Motion**
      1.3.6.1 Types of Forces
      1.3.6.2 Time, Space, and Motion
      1.3.6.3 Dynamics of Motion
      1.3.6.4 Relativity Theory
      1.3.6.5 Fluid Behavior
1.4 **Science, Technology, and Mathematics**
   1.4.1 **Nature or Conceptions of Technology**
   1.4.2 **Interactions of Science, Mathematics, and Technology**
      1.4.2.1 Influence of Mathematics, Technology in Science

                1.4.2.2 Applications of Science in Mathematics, Technology
        1.4.3 **Interactions of Science, Technology, and Society**
                1.4.3.1 Influence of Science, Technology on Society
                1.4.3.2 Influence of Society on Science, Technology
  1.5 **History of Science and Technology**
  1.6 **Environmental and Resource Issues Related to Science**
        1.6.1 **Pollution**
        1.6.2 **Conservation of Land, Water, and Sea Resources**
        1.6.3 **Conservation of Material and Energy Resources**
        1.6.4 **World Population**
        1.6.5 **Food Production, Storage**
        1.6.6 **Effects of Natural Disasters**
  1.7 **Nature of Science**
        1.7.1 **Nature of Scientific Knowledge**
        1.7.2 **The Scientific Enterprise**
  1.8 **Science and Other Disciplines**
        1.8.1 **Science and Mathematics**
        1.8.2 **Science and other Disciplines**

Performance Expectations
  2.1 **Understanding**
        2.1.1 **Simple Information**
        2.1.2 **Complex Information**
        2.1.3 **Thematic Information**
  2.2 **Theorizing, Analyzing, and Solving Problems**
        2.2.1 **Abstracting and Deducing Scientific Principles**
        2.2.2 **Applying Scientific Principles to Solve Quantitative Problems**
        2.2.3 **Applying Scientific Principles to Develop Explanations**
        2.2.4 **Constructing, Interpreting, and Applying Models**
        2.2.5 **Making Decisions**
  2.3 **Using Tools, Routine Procedures, and Science Processes**
        2.3.1 **Using Apparatus, Equipment, and Computers**
        2.3.2 **Conducting Routine Experimental Operations**
        2.3.3 **Gathering Data**
        2.3.4 **Organizing and Representing Data**
        2.3.5 **Interpreting Data**
  2.4 **Investigating the Natural World**
        2.4.1 **Identifying Questions to Investigate**
        2.4.2 **Designing Investigations**

          2.4.3 Conducting Investigations
          2.4.4 Interpreting Investigational Data
          2.4.5 Formulating Conclusions from Investigational Data
2.5 Communicating
          2.5.1 Accessing and Processing Information
          2.5.2 Sharing Information

Perspectives
3.1 Attitudes Toward Science, Mathematics, and Technology
          3.1.1 Positive Attitudes Toward Science, Mathematics, and Technology
          3.1.2 Skeptical Attitudes Toward use of Science and Technology
3.2 Careers in Science, Mathematics and Technology
          3.2.1 Promoting Careers in Science, Mathematics, and Technology
          3.2.2 Promoting Importance of Science, Mathematics, and Technology in Non-technical Careers
3.3 Participation in Science and Mathematics by Underrepresented Groups
3.4 Science, Mathematics, and Technology to Increase Interest
3.5 Safety in Science Performance
3.6 Scientific Habits of Mind

APPENDIX B

Third International Mathematics and Science Study 1995: Teacher (Implemented Curriculum) OTL Questionnaire for Grade 8

Identification Label	
School ID:	
Stratum ID:	
Teacher ID:	Link:
Name:	
Class ID:	
Name of Class:	
Subject:	Grade:

IEA Third International Mathematics and Science Study

Teacher Questionnaire (Mathematics)
Population 2

Your school has agreed to participate in the Third International Mathematics and Science Study (TIMSS), an educational research project sponsored by the International Association for the Evaluation of Educational Achievement (IEA). TIMSS is investigating mathematics and science achievement in over fifty educational systems around the world. It is designed to measure and interpret differences in national educational systems in order to help improve the teaching and learning of mathematics and science worldwide.

This questionnaire is addressed to teachers of mathematics, who are asked to supply information about their academic and professional backgrounds, instructional practices, and attitudes towards teaching mathematics. Since your class has been selected as part of a nationwide sample, your responses are very important in helping to describe mathematics classes in <country>.

Some of the questions in this questionnaire ask about **your mathematics class**. This is the class which is identified at the top of this page, and which will be tested as part of TIMSS in your school.

It is important that you answer each question carefully so that the information provided reflects your situation as accurately as possible. It is estimated that it will require approximately 60 minutes to complete this questionnaire.

Your cooperation in completing this questionnaire is greatly appreciated.

TIMSS Study Center
Boston College   (Institute Address)
Chestnut Hill, MA 02167
USA

Doc. Ref.: ICC880/NRC417
Copyright©IEA, The Hague (1994)

Doc.Ref.: ICC880/NRC417

Mathematics Topics

On the following pages there is a list of mathematics topics. Each topic is illustrated by a short list of subtopics. Not all topics are necessarily appropriate for your class. Nevertheless, please respond to the entire list so that we may obtain an indication of topics covered in your class that is as complete and accurate as possible.

- *Before marking anything, read quickly through the entire list to obtain an idea of where various topics may be found. Be sure to read the four examples on the next page.*

- *If you **have taught** a topic to your class, check the appropriate box indicating the total number of <periods> in which the topic was taught. Four choices are provided: 1-5 <periods>, 6-10 <periods>, 11-15 <periods>, and >15 (i.e., more than 15) <periods>.*

- *If you **will continue to teach or begin teaching** a topic in future lessons this year, check the box in the "will teach later this year" column.*

- *If you have **not taught** a topic and will not teach it this year to your class, check the box in the "not taught this year" column.*

- *If you know that a topic was taught to your students in a **previous year**, check the box in the "taught in a previous year" column.*

- *If you have taught ANY of the subtopics listed under a major topic, indicate that you have taught that major topic area. Subtopics are listed for illustration purposes.*

- *For a few main topics, you are asked to indicate whether you have taught certain subtopics as well as the main topic, since these subtopics are of special interest in this study.*

Doc.Ref.: ICC880/NRC417

EXAMPLES:

NRC Note: <Use country-specific appropriate designation for class <period/hour>.

**How long did you spend teaching each of these topics to your class this year?
Will you cover any of these topics in future <periods>?**

Check as many boxes as apply for each topic listed.

	have taught this year <period> completed				will teach later this year	not taught this year	taught a previous year
	1–5	6–10	11–15	>15			

Example 1. You have not taught this topic and will not teach it this year:

a) **Sets & Logic** .... ☐ ☐ ☐ ☐ | ☐ ☒ ☐
   Sets, set notation and set operations; classification; logic and truth tables

Example 2. You've taught this topic in 2 class <periods> and know it was taught in a previous year:

b) **Problem Solving Strategies** .... ☒ ☐ ☐ ☐ | ☐ ☐ ☒
   Problem solving heuristics and strategies

Example 3. You've taught this topic in 8 class <periods> and will teach it in future <periods>:

c) **Percentages** .... ☐ ☒ ☐ ☐ | ☒ ☐ ☐
   Concepts of percentage; computations with percentage; types of percentage problems

Example 4. You have not taught this topic but will teach it in future <periods>:

d) **Estimation & Error of Measurements** ... ☐ ☐ ☐ ☐ | ☒ ☐ ☐
   Estimation of measurements other than perimeter and area; precision and accuracy; errors of measurement

TQM2 - 18

12. How long did you spend teaching each of these topics to your mathematics class this year? Will you cover any of these topics in future <periods>?

Check as many boxes as apply for each topic listed.

TOPIC	have taught this year <period> completed				will teach later this year	not taught this year	taught a previous year
	1–5	6–10	11–15	>15			
a) **Whole Numbers**	☐	☐	☐	☐	☐	☐	☐
Indicate your coverage both at the main topic level and for each of the following subtopics.							
1. Meaning of whole numbers; place value and numeration	☐	☐	☐	☐	☐	☐	☐
2. Operations with and properties of whole numbers	☐	☐	☐	☐	☐	☐	☐
b) **Common & Decimal Fractions**	☐	☐	☐	☐	☐	☐	☐
Indicate your coverage both at the main topic level and for each of the following subtopics.							
1. Meaning, Representation and Uses of Common Fractions	☐	☐	☐	☐	☐	☐	☐
2. Properties of Common Fractions	☐	☐	☐	☐	☐	☐	☐
3. Meaning, Representation and Uses of Decimal Fractions	☐	☐	☐	☐	☐	☐	☐
4. Properties of Decimal Fractions	☐	☐	☐	☐	☐	☐	☐
5. Relationships Between Common and Decimal Fractions	☐	☐	☐	☐	☐	☐	☐
6. Conversion of Equivalent Forms	☐	☐	☐	☐	☐	☐	☐
7. Ordering of Fractions (Common And Decimals)	☐	☐	☐	☐	☐	☐	☐
c) **Percentages**	☐	☐	☐	☐	☐	☐	☐
Concepts of percentage; computations with percentage; types of percentage problems							
d) **Number Sets & Concepts**	☐	☐	☐	☐	☐	☐	☐
Uses, properties, and computations with integers (negative as well as positive), rational numbers (including negative fractions), real numbers complex numbers; number bases other than ten; exponents, roots and radicals							
e) **Number Theory**	☐	☐	☐	☐	☐	☐	☐
Prime and composite numbers; factorizations of whole numbers; greatest common divisors; least common multiples; permutations; combinations; systematic counting of possibilities and so on							

Doc.Ref.: ICC880/NRC417

TOPIC	have taught this year <period> completed				will teach later this year	not taught this year	taught a previous year
	1–5	6–10	11–15	>15			
f) **Estimation & Number Sense** .... Estimating quantity and size; rounding and significant figures, estimating the results of computations (including mental arithmetic and reasonableness of results); scientific notation and orders of magnitude	☐	☐	☐	☐	☐	☐	☐
g) **Measurement Units & Processes** .... Ideas and units of measurement; standard metric units; length, area, volume, capacity, time, money and so on; use of measurement instruments	☐	☐	☐	☐	☐	☐	☐
h) **Estimation & Error of Measurements** ... Estimation of measurements other than perimeter and area; precision and accuracy; errors of measurement	☐	☐	☐	☐	☐	☐	☐
i) **Perimeter, Area, & Volume** .... Perimeter & area of triangles, quadrilaterals, polygons, circles & other two-dimensional shapes; Calculating, estimating, & solving problems involving perimeters and areas; Surface area and volume	☐	☐	☐	☐	☐	☐	☐
j) **Basics of One & Two Dimensional Geometry** .... Number lines and graphs in one and two dimensions; triangles, quadrilaterals, other polygons, and circles; equations of straight lines; Pythagorean Theorem	☐	☐	☐	☐	☐	☐	☐
k) **Geometric Congruence & Similarity** ..... Concepts, properties and uses of congruent and similar figures, especially for triangles, quadrilaterals, other polygons and plane shapes	☐	☐	☐	☐	☐	☐	☐
l) **Geometric Transformations & Symmetry** Geometric patterns; tessellations; kinds of symmetry in geometric figures, symmetry of number patterns; transformations of all types and their representations; algebraic structure and properties of sets of transformations	☐	☐	☐	☐	☐	☐	☐
m) **Constructions & Three Dimensional Geometry** .... Constructions with compass and straightedge; conic sections; three-dimensional shapes, surfaces and their properties; lines and planes in space; spatial perception and visualization; coordinate graphs and vectors in three dimensions	☐	☐	☐	☐	☐	☐	☐

TOPIC	have taught this year <period> completed				will teach later this year	not taught this year	taught a previous year
	1–5	6–10	11–15	>15			
n) **Ratio & Proportion**	☐	☐	☐	☐	☐	☐	☐
Indicate your coverage both at the main topic level and for each of the following subtopics.							
1. Concepts and Meaning	☐	☐	☐	☐	☐	☐	☐
2. Applications and Uses	☐	☐	☐	☐	☐	☐	☐
Maps and models; solving practical problems based on proportionality; solving proportional equations							
o) **Proportionality: Slope, Trigonometry & Interpolation**	☐	☐	☐	☐	☐	☐	☐
Indicate your coverage both at the main topic level and for each of the following subtopics.							
1. Slope and Trigonometry	☐	☐	☐	☐	☐	☐	☐
Slope; trigonometric ratios; solving triangles and problems involving triangles including the rules of sines and of cosines							
2. Linear Interpolation and Extrapolation .	☐	☐	☐	☐	☐	☐	☐
p) **Functions, Relations, & Patterns**	☐	☐	☐	☐	☐	☐	☐
Number patterns; relations, their properties and graphs; types of function (linear, quadratic, exponential, trigonometric, inverse, etc.); operations on functions; relations of functions and equations (roots, zeros, etc.); problems involving functions							
q) **Equations, Inequalities, & Formulas**	☐	☐	☐	☐	☐	☐	☐
Indicate your coverage both at the main topic level and for each of the following subtopics.							
1. Linear Equations and Formulas	☐	☐	☐	☐	☐	☐	☐
Representing situations algebraically; work with formulas other than measurement formulas; algebraic expressions & working with them (Factoring, polynomial operations, etc.); solving linear equations							
2. Other Equations and Formulas	☐	☐	☐	☐	☐	☐	☐
Solving various types of equations (quadratic, radical, trigonometric, logarithmic, etc.); inequalities; systems of equations; systems of inequalities							

Doc.Ref.: ICC880/NRC417

TOPIC	have taught this year <period> completed				will teach later this year	not taught this year	taught a previous year
	1–5	6–10	11–15	>15			
r) **Statistics & Data** .... Collecting data from experiments & surveys; representing & interpreting data in tables, charts, graphs, etc.; nominal, ordinal, etc., scales; means, medians & other measures of central tendency; variance, standard deviations & other measure of dispersion; sampling, randomness & bias; prediction & inferences from data; regression & fitting lines & curves to data; correlation's & other measures of relationship; use & misuse of statistics in analyzing data	☐	☐	☐	☐	☐	☐	☐
s) **Probability & Uncertainty** .... Informal language of 'more likely,' 'less likely', etc.; probability models & numerical probability; all other aspects of probability & probability distributions for random variables; expectations, parameter estimation, hypothesis testing, confidence intervals, & related statistical topics	☐	☐	☐	☐	☐	☐	☐
t) **Sets & Logic** .... Sets, set notation and set operations; classification; logic and truth tables	☐	☐	☐	☐	☐	☐	☐
u) **Problem Solving Strategies** .... Problem solving heuristics and strategies	☐	☐	☐	☐	☐	☐	☐
v) **Other Mathematics Content** .... Mark here for all content you covered that was not in one of the earlier categories. This includes advanced topics such as the following: Computers (operation of computers, flow charts, learning a programming language, programs, algorithms with applications to the computer); History and nature of mathematics; and Proofs.	☐	☐	☐	☐	☐	☐	☐

TQM2 - 22

Think of the last <lesson> in which you taught mathematics to your mathematics class. (If this lesson was atypical, e.g. an examination period or a field trip, pick the previous one.)

13a. **How many minutes was this class <hour/period>?**

Please write in a number.

_____ minutes

13b. **For each of the following mathematics topics, indicate whether or not it was the subject of the lesson.**

(See 'Mathematics Topics' category descriptions in question 12.)

Check one box in each row.

	Yes	No
1. Whole Numbers	☐	☐
2. Common and Decimal Fractions	☐	☐
3. Percentages	☐	☐
4. Number Sets and Concepts	☐	☐
5. Number Theory	☐	☐
6. Estimation and Number Sense	☐	☐
7. Measurement Units and Processes	☐	☐
8. Estimation and Error of Measurements	☐	☐
9. Perimeter, Area and Volume	☐	☐
10. Basics of One and Two Dimensional Geometry	☐	☐
11. Geometric Congruence and Similarity	☐	☐
12. Geometric Transformations and Symmetry	☐	☐
13. Constructions and Three Dimensional Geometry	☐	☐
14. Ratio and Proportion	☐	☐
15. Proportionality: Slope, trigonometry and interpolation	☐	☐
16. Functions, Relations, and Patterns	☐	☐
17. Equations, Inequalities, and Formulas	☐	☐
18. Statistics and Data	☐	☐
19. Probability and Uncertainty	☐	☐
20. Sets and Logic	☐	☐
21. Problem Solving Strategies	☐	☐
22. Other Mathematics Content	☐	☐

13c. **Was this lesson...**

Check one box in each row.

	Yes	No
1. the introduction of this topic	☐	☐
2. a continuation of a previous lesson on the same topic	☐	☐
3. the end of the coverage of this topic	☐	☐

Identification Label	
School ID :	
Stratum ID:	
Teacher ID:	Link:
Name:	
Class ID:	
Name of Class:	
Subject:	Grade:

IEA Third International Mathematics and Science Study

Teacher Questionnaire (Science)
Population 2

Your school has agreed to participate in the Third International Mathematics and Science Study (TIMSS), an educational research project sponsored by the International Association for the Evaluation of Educational Achievement (IEA). TIMSS is investigating mathematics and science achievement in over fifty educational systems around the world. It is designed to measure and interpret differences in national educational systems in order to help improve the teaching and learning of mathematics and science worldwide.

This questionnaire is addressed to teachers of science, who are asked to supply information about their academic and professional backgrounds, instructional practices, and attitudes towards teaching science. Since your class has been selected as part of a nationwide sample, your responses are very important in helping to describe science classes in <country>.

Some of the questions in this questionnaire ask about **your science class**. This is the class which is identified at the top of this page, and which will be tested as part of TIMSS in your school.

It is important that you answer each question carefully so that the information provided reflects your situation as accurately as possible. It is estimated that it will require approximately 60 minutes to complete this questionnaire.

Your cooperation in completing this questionnaire is greatly appreciated.

TIMSS Study Center
Boston College   (Institute Address)
Chestnut Hill, MA 02167
USA

Doc. Ref.: ICC881/NRC418
Copyright©IEA, The Hague (1994)

Doc.Ref.: ICC881/NRC418

Science Topics

On the following pages there is a list of science topics. Each topic is illustrated by a short list of subtopics. Not all topics are necessarily appropriate for your class. Nevertheless, please respond to the entire list so that we may obtain an indication of topics covered in your class that is as complete and accurate as possible.

- *Before marking anything, read quickly through the entire list to obtain an idea of where various topics may be found. Be sure to read the four examples on the next page.*

- *If you **have taught** a topic to your class, check the appropriate box indicating the total number of <periods> in which the topic was taught. Four choices are provided: 1-5 <periods>, 6-10 <periods>, 11-15 <periods>, and >15 (i.e., more than 15) <periods>.*

- *If you **will continue to teach or begin teaching** a topic in future lessons this year, check the box in the "will teach later this year" column.*

- *If you have **not taught** a topic and will not teach it this year to your class, check the box in the "not taught this year" column.*

- *If you know that a topic was taught to your students in a **previous year**, check the box in the "taught in a previous year" column.*

- *If you have taught ANY of the subtopics listed under a major topic, indicate that you have taught that major topic area. Subtopics are listed for illustration purposes.*

- *For a few main topics, you are asked to indicate whether you have taught certain subtopics as well as the main topic, since these subtopics are of special interest in this study.*

Doc.Ref.: ICC881/NRC418

EXAMPLES:

NRC Note: <Use country-specific appropriate designation for class <period/hour>.

How long did you spend teaching each of these topics to your class this year?
Will you cover any of these topics in future <periods>?

Check as many boxes as apply for each topic listed.

	have taught this year <period> completed				will teach later this year	not taught this year	taught a previous year
	1–5	6–10	11–15	>15			

Example 1. You have not taught this topic and will not teach it this year:

a) **Relativity Theory** ....
   Relativity theory
   □ □ □ □ | □ ☒ □

Example 2. You've taught this topic in 2 class <periods> and know it was taught in a previous year:

b) **Earth Processes** ....
   Weather and climate, physical cycles, building and breaking (e.g., volcanoes, earthquakes), geologic timetable, fossils
   ☒ □ □ □ | □ □ ☒

Example 3. You've taught this topic in 8 class <periods> and will teach it in future <periods>:

c) **Energy Processes** ....
   Heat and temperature; wave phenomena, sound and vibration, electricity, and magnetism
   □ □ ☒ □ | □ □ □

Example 4. You have not taught this topic but will teach it in future <periods>:

d) **Nature of Science** ....
   The nature of scientific knowledge; the scientific enterprise; and scientific methods
   □ □ □ □ | ☒ □ □

12. How long did you spend teaching each of these topics to your science class <u>this year</u>? Will you cover any of these topics in future <periods>?

Check as many boxes as apply for each topic listed.

TOPIC	have taught this year <period> completed				will teach later this year	not taught this year	taught a previous year
	1–5	6–10	11–15	>15			
a) **Earth Features**	☐	☐	☐	☐	☐	☐	☐
Indicate your coverage both at the main topic level and for each of the following subtopics.							
1. Layers of the Earth	☐	☐	☐	☐	☐	☐	☐
2. Landforms	☐	☐	☐	☐	☐	☐	☐
3. Bodies of Water	☐	☐	☐	☐	☐	☐	☐
4. Atmosphere	☐	☐	☐	☐	☐	☐	☐
5. Rocks, Soil	☐	☐	☐	☐	☐	☐	☐
6. Iceforms	☐	☐	☐	☐	☐	☐	☐
b) **Earth Processes**	☐	☐	☐	☐	☐	☐	☐
Weather and climate, physical cycles, building and breaking (e.g., volcanoes, earthquakes), geologic timetable, fossils							
c) **Earth in the Universe**	☐	☐	☐	☐	☐	☐	☐
Interactions between sun, earth and moon; planets and the solar system; things beyond the solar system; evolution of the universe							
d) **Human Biology & Health**	☐	☐	☐	☐	☐	☐	☐
1. Structures and Functions of the Body	☐	☐	☐	☐	☐	☐	☐
2. Metabolism, Respiration, Digestion and other Bodily Processes	☐	☐	☐	☐	☐	☐	☐
3. Reproduction	☐	☐	☐	☐	☐	☐	☐
4. Genetics	☐	☐	☐	☐	☐	☐	☐
e) **Diversity & Structure of Living Things**	☐	☐	☐	☐	☐	☐	☐
Plants, fungi, animals, other organisms; structure and function of organ systems; organs, tissues, and cells							
f) **Life Processes & Systems Enabling Life Functions**	☐	☐	☐	☐	☐	☐	☐
Sensing and responding; biochemical processes in cells; photosynthesis; respiration; digestion							
g) **Life Cycles, Genetic Continuity, Diversity**	☐	☐	☐	☐	☐	☐	☐
Life cycles, reproduction, variation and inheritance, evolution, speciation, diversity, and the biochemistry of genetics							

TOPIC	have taught this year <period> completed				will teach later this year	not taught this year	taught a previous year
	1-5	6-10	11-15	>15			
h) **Interactions of Living Things**............ Biomes and ecosystems, habitats and niches; the interdependence of life; animal behavior'	☐	☐	☐	☐	☐	☐	☐
i) **Types and Properties of Matter**............ Classification of matter (e.g., mixtures, compounds); physical properties and chemical properties	☐	☐	☐	☐	☐	☐	☐
j) **Structure of Matter**............................. Atoms, ions, molecules, macromolecules, crystals	☐	☐	☐	☐	☐	☐	☐
k) **Energy Types, Sources, and Conversions** Types of energy (e.g., mechanical, chemical); sources of energy (food, oil, wood); conversions of energy; work and efficiency	☐	☐	☐	☐	☐	☐	☐
l) **Energy Processes**............................... Heat and temperature; wave phenomena, sound and vibration, electricity, and magnetism. *Indicate your coverage both for the above main topic and for the following subtopic.*	☐	☐	☐	☐	☐	☐	☐
1. Light..	☐	☐	☐	☐	☐	☐	☐
m) **Physical Changes**............................. Physical changes and explanations of physical changes	☐	☐	☐	☐	☐	☐	☐
n) **Kinetic & Quantum Theory**.............. Kinetic theory and quantum theory and fundamental particles	☐	☐	☐	☐	☐	☐	☐
o) **General Chemical Changes**............... Chemical changes, explanations of chemical changes; rate of change and equilibria; energy and chemical change	☐	☐	☐	☐	☐	☐	☐
p) **Specialized Chemical Changes**............ Nuclear fusion and fission; radiation; electrochemistry; organic and biochemical changes	☐	☐	☐	☐	☐	☐	☐
q) **Forces & Motion**............................. Types of forces; speed, acceleration; dynamics of motion, fluid behavior	☐	☐	☐	☐	☐	☐	☐
r) **Relativity Theory**............................. Relativity theory	☐	☐	☐	☐	☐	☐	☐

TQS2 - 20

TOPIC	have taught this year <period> completed				will teach later this year	not taught this year	taught a previous year
	1-5	6-10	11-15	>15			
s) **Science, Technology, & Society** The nature or conceptions of technology; the interactions among science, mathematics, and technology, and the interactions between science, technology and society	☐	☐	☐	☐	☐	☐	☐
t) **History of Science & Technology** Famous scientists, classic experiments, historical development of scientific ideas, the industrial revolution, and classic inventions	☐	☐	☐	☐	☐	☐	☐
u) **Environmental & Resource Issues** Pollution, conservation of land, water, and sea resources, conservation of material and energy resources, world population, food production and storage, and the effects of natural disasters	☐	☐	☐	☐	☐	☐	☐
v) **Nature of Science** Enterprise; and scientific methods	☐	☐	☐	☐	☐	☐	☐
w) **Measurement** 1. Using Apparatus	☐	☐	☐	☐	☐	☐	☐
2. Conducting Routine Experimental Operations	☐	☐	☐	☐	☐	☐	☐
3. Gathering Data	☐	☐	☐	☐	☐	☐	☐
x) **Data Analysis** 1. Organizing and Representing Data	☐	☐	☐	☐	☐	☐	☐
2. Interpreting Provided Data	☐	☐	☐	☐	☐	☐	☐
3. Interpreting Data From Student Investigations	☐	☐	☐	☐	☐	☐	☐
4. Formulating Conclusions From Data Collected	☐	☐	☐	☐	☐	☐	☐
By Students	☐	☐	☐	☐	☐	☐	☐

Doc.Ref.: ICC881/NRC418

Think of the last <lesson> in which you taught science to your science class. (If this lesson was atypical, e.g., an examination period or a field trip, pick the previous one.)

13a. How many minutes was this class <hour/period>?

Please write in a number.

_____ minutes

13b. For each of the following science topics, indicate whether or not it was the subject of the lesson.
(See "Science Topics" category descriptions in question 12.)

Check one box in each row.

	Yes	No
1. Earth Features	☐	☐
2. Earth Processes	☐	☐
3. Earth in the Universe	☐	☐
4. Human Biology and Health	☐	☐
5. Diversity and Structure of Living Things	☐	☐
6. Life Processes and Systems Enabling Life Function	☐	☐
7. Life Cycles, Genetic Continuity, Diversity	☐	☐
8. Interactions of Living Things	☐	☐
9. Types and Properties of Matter	☐	☐
10. Structure of Matter	☐	☐
11. Energy Types, Sources, and Conversions	☐	☐
12. Energy Processes	☐	☐
13. Physical Changes	☐	☐
14. Kinetic and Quantum Theory	☐	☐
15. General Chemical Changes	☐	☐
16. Specialized Chemical Changes	☐	☐
17. Force and Motion	☐	☐
18. Relativity Theory	☐	☐
19. Science, Technology and Society	☐	☐
20. History of Science and Technology	☐	☐
21. Environmental and Resource Issues	☐	☐
22. Nature of Science	☐	☐

13c. Was this lesson...

Check one box in each row.

	Yes	No
1. the introduction of this topic	☐	☐
2. a continuation of a previous lesson on the same topic	☐	☐
3. the end of the coverage of this topic	☐	☐

APPENDIX C

Programme for International Student Assessment 2012: Student (Implemented Curriculum) OTL Questionnaire

Programme for International Student Assessment 2012

ANNEX A: BACKGROUND QUESTIONNAIRES

Q39 Thinking about mathematical concepts: how familiar are you with the following terms?
(ST62) *(Please tick only one box in each row.)*

		Never heard of it	Heard of it once or twice	Heard of it a few times	Heard of it often	Know it well, understand the concept
a)	Exponential Function	□₁	□₂	□₃	□₄	□₅
b)	Divisor	□₁	□₂	□₃	□₄	□₅
c)	Quadratic Function	□₁	□₂	□₃	□₄	□₅
d)	<Proper Number>	□₁	□₂	□₃	□₄	□₅
e)	Linear Equation	□₁	□₂	□₃	□₄	□₅
f)	Vectors	□₁	□₂	□₃	□₄	□₅
g)	Complex Number	□₁	□₂	□₃	□₄	□₅
h)	Rational Number	□₁	□₂	□₃	□₄	□₅
i)	Radicals	□₁	□₂	□₃	□₄	□₅
j)	<Subjunctive Scaling>	□₁	□₂	□₃	□₄	□₅
k)	Polygon	□₁	□₂	□₃	□₄	□₅
l)	<Declarative Fraction>	□₁	□₂	□₃	□₄	□₅
m)	Congruent Figure	□₁	□₂	□₃	□₄	□₅
n)	Cosine	□₁	□₂	□₃	□₄	□₅
o)	Arithmetic Mean	□₁	□₂	□₃	□₄	□₅
p)	Probability	□₁	□₂	□₃	□₄	□₅

Q40 How many minutes, on average, are there in a <class period> for the following subjects?
(ST69)

		Minutes
a)	Minutes in a <class period> in <test language>	
b)	Minutes in a <class period> in mathematics	
c)	Minutes in a <class period> in <science>	

Notes: <**Class period**> refers to the length of time each lesson runs for in a normal school week.
For a definition of <**test language**> and <**science**> see Q33.

Q41 How many <class periods> per week do you typically have for the following subjects?
(ST70)

		<class periods>
a)	Number of <class periods> per week in <test language>	
b)	Number of <class periods> per week in mathematics	
c)	Number of <class periods> per week in <science>	

Notes: For a definition of <**class period**> see Q40.
For a definition of <**test language**> and <**science**> see Q33.

ANNEX A: BACKGROUND QUESTIONNAIRES

In the next type of problem, you have to use mathematical knowledge and draw conclusions. There is no practical application provided. Here are two examples.

1) Here you need to use geometrical theorems:

Determine the height of the pyramid

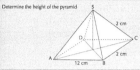

Q46
(ST75)

2) Here you have to know what a prime number is:

If n is any number: can $(n+1)^2$ be a prime number?

We want to know about your experience with these types of problems at school.
Do not solve them!

(Please tick only one box in each row.)

		Frequently	Sometimes	Rarely	Never
a)	How often have you encountered these types of problems in your **mathematics lessons**?	☐₁	☐₂	☐₃	☐₄
b)	How often have you encountered these types of problems in the **tests you have taken at school**?	☐₁	☐₂	☐₃	☐₄

In this type of problem, you have to apply suitable mathematical knowledge to find a useful answer to a problem that arises in everyday life or work. The data and information are about real situations. Here are two examples.

Example 1:

A TV reporter says "This graph shows that there is a huge increase in the number of robberies from 1998 to 1999."

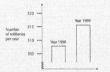

Q47
(ST76)

Do you consider the reporter's statement to be a reasonable interpretation of the graph?
Give an explanation to support your answer.

Example 2:

For years the relationship between a person's recommended maximum heart rate and the person's age was described by the following formula:
Recommended maximum heart rate = 220 - age
Recent research showed that this formula should be modified slightly. The new formula is as follows:
Recommended maximum heart rate = 208 − (0.7 × age)
From which age onwards does the recommended maximum heart rate increase as a result of the introduction of the new formula? Show your work.

We want to know about your experience with these types of problems at school.
Do not solve them!

(Please tick only one box in each row.)

		Frequently	Sometimes	Rarely	Never
a)	How often have you encountered these types of problems in your **mathematics lessons**?	☐₁	☐₂	☐₃	☐₄
b)	How often have you encountered these types of problems in the **tests you have taken at school**?	☐₁	☐₂	☐₃	☐₄

APPENDIX D

Trends in International Mathematics and Science Study 2015: Teacher (Implemented Curriculum) OTL Questionnaire for Grade 8 and Country Expert (Intended Curriculum) OTL Questionnaire

TRENDS IN INTERNATIONAL MATHEMATICS AND SCIENCE STUDY

Teacher Questionnaire Mathematics

<Grade 8>
<TIMSS National Research Center Name>
<Address>

© IEA, 2014

Mathematics Topics Taught to the TIMSS Class

21

The following list includes the main topics addressed by the TIMSS mathematics test. Choose the response that best describes when the students in this class have been taught each topic. If a topic was in the curriculum before the <eighth grade>, please choose "Mostly taught before this year." If a topic was taught half this year but not yet completed, please choose "Mostly taught this year." If a topic is not in the curriculum, please choose "Not yet taught or just introduced."

Check **one** circle for each line.

	Mostly taught before this year	Mostly taught this year	Not yet taught or just introduced
A. Number			
a) Computing with whole numbers	○	○	○
b) Comparing and ordering rational numbers	○	○	○
c) Computing with rational numbers (fractions, decimals, and integers)	○	○	○
d) Concepts of irrational numbers	○	○	○
e) Problem solving involving percents or proportions	○	○	○
B. Algebra			
a) Simplifying and evaluating algebraic expressions	○	○	○
b) Simple linear equations and inequalities	○	○	○
c) Simultaneous (two variables) equations	○	○	○
d) Numeric, algebraic, and geometric patterns or sequences (extension, missing terms, generalization of patterns)	○	○	○
e) Representation of functions as ordered pairs, tables, graphs, words, or equations	○	○	○
f) Properties of functions (slopes, intercepts, etc.)	○	○	○
C. Geometry			
a) Geometric properties of angles and geometric shapes (triangles, quadrilaterals, and other common polygons)	○	○	○
b) Congruent figures and similar triangles	○	○	○
c) Relationship between three-dimensional shapes and their two-dimensional representations	○	○	○
d) Using appropriate measurement formulas for perimeters, circumferences, areas, surface areas, and volumes	○	○	○
e) Points on the Cartesian plane	○	○	○
f) Translation, reflection, and rotation	○	○	○
D. Data and Chance			
a) Characteristics of data sets (mean, median, mode, and shape of distributions)	○	○	○
b) Interpreting data sets (e.g., draw conclusions, make predictions, and estimate values between and beyond given data points)	○	○	○
c) Judging, predicting, and determining the chances of possible outcomes	○	○	○

TRENDS IN INTERNATIONAL MATHEMATICS AND SCIENCE STUDY

Teacher Questionnaire Science

<Grade 8>
<TIMSS National Research Center Name>
<Address>

© IEA, 2014

Science Topics Taught to the <TIMSS Class/Class with the TIMSS students>

20 The following list includes the main topics addressed by the TIMSS science test. Choose the response that best describes when the students in this class have been taught each topic. If a topic was in the curriculum before the <eighth grade>, please choose "Mostly taught before this year." If a topic was taught half this year but not yet completed, please choose "Mostly taught this year." If a topic is not in the curriculum, please choose "Not yet taught or just introduced."

Check one circle for each line.

	Mostly taught before this year	Mostly taught this year	Not yet taught or just introduced

A. Biology

a) Differences among major taxonomic groups of organisms (plants, animals, fungi, mammals, birds, reptiles, fish, amphibians) — ○ — ○ — ○

b) Major organs and organ systems in humans and other organisms (structure/function, life processes that maintain stable bodily conditions) — ○ — ○ — ○

c) Cells, their structure and functions, including respiration and photosynthesis as cellular processes — ○ — ○ — ○

d) Life cycles, sexual reproduction, and heredity (passing on of traits, inherited versus acquired/learned characteristics) — ○ — ○ — ○

e) Role of variation and adaptation in survival/extinction of species in a changing environment (including fossil evidence for changes in life on Earth over time) — ○ — ○ — ○

f) Interdependence of populations of organisms in an ecosystem (e.g., energy flow, food webs, competition, predation) and factors affecting population size in an ecosystem — ○ — ○ — ○

g) Human health (causes of infectious diseases, methods of infection, prevention, immunity) and the importance of diet and exercise in maintaining health — ○ — ○ — ○

B. Chemistry

a) Classification, composition, and particulate structure of matter (elements, compounds, mixtures, molecules, atoms, protons, neutrons, electrons) — ○ — ○ — ○

b) Physical and chemical properties of matter — ○ — ○ — ○

c) Mixtures and solutions (solvent, solute, concentration/dilution, effect of temperature on solubility) — ○ — ○ — ○

d) Properties and uses of common acids and bases — ○ — ○ — ○

e) Chemical change (transformation of reactants, evidence of chemical change, conservation of matter, common oxidation reactions – combustion, rusting, tarnishing) — ○ — ○ — ○

f) The role of electrons in chemical bonds — ○ — ○ — ○

20 (continued)

Choose the response that best describes when the students in this class have been taught each topic. If a topic was in the curriculum before the <u>eighth grade</u>, please choose "Mostly taught before this year." If a topic was taught half this year but not yet completed, please choose "Mostly taught this year." If a topic is not in the curriculum, please choose "Not yet taught or just introduced."

Check **one** circle for each line.

	Mostly taught before this year	Mostly taught this year	Not yet taught or just introduced
C. Physics			
a) Physical states and changes in matter (explanations of properties in terms of movement and distance between particles; phase change, thermal expansion, and changes in volume and/or pressure)	○	○	○
b) Energy forms, transformations, heat, and temperature	○	○	○
c) Basic properties/behaviors of light (reflection, refraction, light and color, simple ray diagrams) and sound (transmission through media, loudness, pitch, amplitude, frequency)	○	○	○
d) Electric circuits (flow of current; types of circuits - parallel/series) and properties and uses of permanent magnets and electromagnets	○	○	○
e) Forces and motion (types of forces, basic description of motion, effects of density and pressure)	○	○	○
D. Earth Science			
a) Earth's structure and physical features (Earth's crust, mantle, and core; composition and relative distribution of water, and composition of air)	○	○	○
b) Earth's processes, cycles, and history (rock cycle; water cycle; weather versus climate; major geological events; formation of fossils and fossil fuels)	○	○	○
c) Earth's resources, their use and conservation (e.g., renewable/nonrenewable resources, human use of land/soil, water resources)	○	○	○
d) Earth in the solar system and the universe (phenomena on Earth - day/night, tides, phases of moon, eclipses, seasons; physical features of Earth compared to other bodies)	○	○	○

<Grade 8> Teacher *Questionnaire* – *Science*

TIMSS 2015 Curriculum Questionnaire— Eighth Grade

TIMSS - 2015 - English
You are logged in as: 9911 Logout

TIMSS 2015 Curriculum Questionnaire – Eighth Grade - Eighth Grade Mathematics Topics Covered

Eighth Grade Mathematics Topics Covered

This mathematics module refers to the national curriculum that was in effect for the eighth grade students assessed in TIMSS 2015—the curriculum that covers mathematics instruction at the eighth grade of formal schooling for the majority of students. If you do not have a national curriculum, please summarize for your state or provincial curricula.

M8. (i) According to the national mathematics curriculum, what proportion of grade 8 students should have been taught each of the following topics or skills by the end of grade 8?

Be sure to include curriculum expectations for all grades up to and including grade 8. Grades represent years of formal schooling. For example, if "Year 9" in your country corresponds to the eighth year of formal schooling, please choose grade 8.

(ii) Across grades from preprimary through upper secondary education, at what grade(s) are the topics primarily intended to be taught?

If there are not any specifications to this detail, please indicate national expectations to the best of your ability. If part of a topic does not apply (e.g., fractions in part A topic (c)), please explain in the comment field.

	(i) Proportion of grade 8 students expected to be taught topic			(ii) Grade(s) topic is expected to be taught preprimary (PP) through the end of upper secondary (G12)											
	Check one circle for each line.			Check the corresponding grade(s) for each topic.											
A. Number	All or almost all students	Only the more able students	Not included in the curriculum through grade 8	PP	G1	G2	G3	G4	G5	G6	G7	G8	G9	G10	G11	G12
a) Computing with whole numbers	○	○	○	○	○	○	○	○	○	○	○	○	○	○	○	○
b) Comparing and ordering rational numbers	○	○	○	○	○	○	○	○	○	○	○	○	○	○	○	○
c) Computing with rational numbers (fractions, decimals, and integers)	○	○	○	○	○	○	○	○	○	○	○	○	○	○	○	○
d) Concepts of irrational numbers	○	○	○	○	○	○	○	○	○	○	○	○	○	○	○	○
e) Problem solving involving percents or proportions	○	○	○	○	○	○	○	○	○	○	○	○	○	○	○	○

Comments:

Previous   25/40   Table of Contents   Next

© IEA Online SurveySystem 2015 - Help

TIMSS - 2015 - English
You are logged in as: 9911 Logout

TIMSS 2015 Curriculum Questionnaire – Eighth Grade - Eighth Grade Mathematics Topics Covered

M8. (continued)
(i) According to the national mathematics curriculum, what proportion of grade 8 students should have been taught each of the following topics or skills by the end of grade 8?

Be sure to include curriculum expectations for all grades up to and including grade 8. Grades represent years of formal schooling. For example, if "Year 9" in your country corresponds to the eighth year of formal schooling, please choose grade 8.

(ii) Across grades from preprimary through upper secondary education, at what grade(s) are the topics primarily intended to be taught?

If there are not any specifications to this detail, please indicate national expectations to the best of your ability. If part of a topic does not apply (e.g., fractions in part A topic (c)), please explain in the comment field.

	(I) Proportion of grade 8 students expected to be taught topic			(ii) Grade(s) topic is expected to be taught preprimary (PP) through the end of upper secondary (G12)											
	Check **one** circle for each line.			Check the corresponding grade(s) for each topic.											
B. Algebra	All or almost all students	Only the more able students	Not included in the curriculum through grade 8	PP	G1	G2	G3	G4	G5	G6	G7	G8	G9	G10	G11	G12
a) Simplifying and evaluating algebraic expressions	○	○	○	○	○	○	○	○	○	○	○	○	○	○	○	○
b) Simple linear equations and inequalities	○	○	○	○	○	○	○	○	○	○	○	○	○	○	○	○
c) Simultaneous (two variables) equations	○	○	○	○	○	○	○	○	○	○	○	○	○	○	○	○
d) Numeric, algebraic, and geometric patterns or sequences (extension, missing terms, generalization of patterns)	○	○	○	○	○	○	○	○	○	○	○	○	○	○	○	○
e) Representation of functions as ordered pairs, tables, graphs, words, or equations	○	○	○	○	○	○	○	○	○	○	○	○	○	○	○	○
f) Properties of functions (slopes, intercepts, etc.)	○	○	○	○	○	○	○	○	○	○	○	○	○	○	○	○

Comments:

[Previous]   26/40   Table of Contents   [Next]

© IEA Online SurveySystem 2015 - Help

TIMSS - 2015 - English
You are logged in as: 9911 Logout

TIMSS 2015 Curriculum Questionnaire – Eighth Grade - Eighth Grade Mathematics Topics Covered

M8. (continued)
(i) According to the national mathematics curriculum, what proportion of grade 8 students should have been taught each of the following topics or skills by the end of grade 8?

Be sure to include curriculum expectations for all grades up to and including grade 8. Grades represent years of formal schooling. For example, if "Year 9" in your country corresponds to the eighth year of formal schooling, please choose grade 8.

(ii) Across grades from preprimary through upper secondary education, at what grade(s) are the topics primarily intended to be taught?

If there are not any specifications to this detail, please indicate national expectations to the best of your ability. If part of a topic does not apply [e.g., fractions in part A topic (c)], please explain in the comment field.

	(i) Proportion of grade 8 students expected to be taught topic			(ii) Grade(s) topic is expected to be taught preprimary (PP) through the end of upper secondary (G12)											
	Check one circle for each line.			Check the corresponding grade(s) for each topic.											
C. Geometry	All or almost all students	Only the more able students	Not Included in the curriculum through grade 8	PP	G1	G2	G3	G4	G5	G6	G7	G8	G9	G10	G11	G12
a) Geometric properties of angles and geometric shapes (triangles, quadrilaterals, and other common polygons)	○	○	○	□	□	□	□	□	□	□	□	□	□	□	□	□
b) Congruent figures and similar triangles	○	○	○	□	□	□	□	□	□	□	□	□	□	□	□	□
c) Relationship between three-dimensional shapes and their two-dimensional representations	○	○	○	□	□	□	□	□	□	□	□	□	□	□	□	□
d) Using appropriate measurement formulas for perimeters, circumferences, areas, surface areas, and volumes	○	○	○	□	□	□	□	□	□	□	□	□	□	□	□	□
e) Points on the Cartesian plane	○	○	○	□	□	□	□	□	□	□	□	□	□	□	□	□
f) Translation, reflection, and rotation	○	○	○	□	□	□	□	□	□	□	□	□	□	□	□	□

Comments:

[Previous]   27/40 Table of Contents   [Next]

© IEA Online SurveySystem 2015 - Help

Grade 8
CURRICULUM QUESTIONNAIRE

Trends in International Mathematics and Science Study 2015

TIMSS - 2015 - English
You are logged in as: 9911 Logout

TIMSS 2015 Curriculum Questionnaire – Eighth Grade - Eighth Grade Mathematics Topics Covered

M8. (continued)
(i) According to the national mathematics curriculum, what proportion of grade 8 students should have been taught each of the following topics or skills by the end of grade 8?

Be sure to include curriculum expectations for all grades up to and including grade 8. Grades represent years of formal schooling. For example, if "Year 9" in your country corresponds to the eighth year of formal schooling, please choose grade 8.

(ii) Across grades from preprimary through upper secondary education, at what grade(s) are the topics primarily intended to be taught?

If there are not any specifications to this detail, please indicate national expectations to the best of your ability. If part of a topic does not apply (e.g., fractions in part A topic (c)], please explain in the comment field.

	(i) Proportion of grade 8 students expected to be taught topic			(ii) Grade(s) topic is expected to be taught preprimary (PP) through the end of upper secondary (G12)											
	Check one circle for each line.			Check the corresponding grade(s) for each topic.											
D. Data and Chance	All or almost all students	Only the more able students	Not included in the curriculum through grade 8	PP	G1	G2	G3	G4	G5	G6	G7	G8	G9	G10	G11	G12
a) Characteristics of data sets (mean, median, mode, and shape of distributions)	○	○	○	○	○	○	○	○	○	○	○	○	○	○	○	○
b) Interpreting data sets (e.g., draw conclusions, make predictions, and estimate values between and beyond given data points)	○	○	○	○	○	○	○	○	○	○	○	○	○	○	○	○
c) Judging, predicting, and determining the chances of possible outcomes	○	○	○	○	○	○	○	○	○	○	○	○	○	○	○	○

Comments:

© IEA Online SurveySystem 2015 - Help

TIMSS - 2015 - English
You are logged in as: 9911 Logout

TIMSS 2015 Curriculum Questionnaire – Eighth Grade - Eighth Grade Science Topics Covered

Eighth Grade Science Topics Covered

This science module refers to the national curriculum that was in effect for the eighth grade students assessed in TIMSS 2015—the curriculum that covers science instruction at the eighth grade of formal schooling for the majority of students. If you do not have a national curriculum, please summarize for your state or provincial curricula.

S8. (i) According to the national science curriculum, what proportion of grade 8 students should have been taught each of the following topics or skills by the end of grade 8?

Be sure to include curriculum expectations for all grades up to and including grade 8. Grades represent years of formal schooling. For example, if "Year 9" in your country corresponds to the eighth year of formal schooling, please choose grade 8.

(ii) Across grades from preprimary through upper secondary education, at what grade(s) are the topics primarily intended to be taught?

If there are not any specifications to this detail, please indicate national expectations to the best of your ability. If part of a topic does not apply (e.g., energy flow in part A topic (f)), please explain in the comment field.

	(i) Proportion of grade 8 students expected to be taught topic			(ii) Grade(s) topic is expected to be taught preprimary (PP) through the end of upper secondary (G12)											
	Check one circle for each line.			Check the corresponding grade(s) for each topic											
A. Biology	All or almost all students	Only the more able students	Not included in the curriculum through grade 8	PP	G1	G2	G3	G4	G5	G6	G7	G8	G9	G10	G11	G12
a) Differences among major taxonomic groups of organisms (plants, animals, fungi, mammals, birds, reptiles, fish, amphibians)	○	○	○	☐	☐	☐	☐	☐	☐	☐	☐	☐	☐	☐	☐	☐
b) Major organs and organ systems in humans and other organisms (structure/function, life processes that maintain stable bodily conditions)	○	○	○	☐	☐	☐	☐	☐	☐	☐	☐	☐	☐	☐	☐	☐
c) Cells, their structure and functions, including respiration and photosynthesis as cellular processes	○	○	○	☐	☐	☐	☐	☐	☐	☐	☐	☐	☐	☐	☐	☐
d) Life cycles, sexual reproduction, and heredity (passing on of traits, inherited versus acquired/learned characteristics)	○	○	○	☐	☐	☐	☐	☐	☐	☐	☐	☐	☐	☐	☐	☐
e) Role of variation and adaptation in survival/extinction of species in a changing environment (including fossil evidence for changes in life on Earth over time)	○	○	○	☐	☐	☐	☐	☐	☐	☐	☐	☐	☐	☐	☐	☐
f) Interdependence of populations of organisms in an ecosystem (e.g., energy flow, food webs, competition, predation) and factors affecting population size in an ecosystem	○	○	○	☐	☐	☐	☐	☐	☐	☐	☐	☐	☐	☐	☐	☐
g) Human health (causes of infectious diseases, methods of infection, prevention, immunity) and the importance of diet and exercise in maintaining health	○	○	○	☐	☐	☐	☐	☐	☐	☐	☐	☐	☐	☐	☐	☐

(Continued on Next Page)

© IEA Online SurveySystem 2015 - Help

Grade 8
CURRICULUM QUESTIONNAIRE

TIMSS - 2015 - English *(Continued)*
You are logged in as: 9911 Logout

TIMSS 2015 Curriculum Questionnaire – Eighth Grade - Eighth Grade Science Topics Covered

Comments:

[Previous] 37/40 Table of Contents [Next]

© IEA Online SurveySystem 2015 - Help

S8. (continued)

(I) According to the national science curriculum, what proportion of grade 8 students should have been taught each of the following topics or skills by the end of grade 8?

Be sure to include curriculum expectations for all grades up to and including grade 8. Grades represent years of formal schooling. For example, if "Year 9" in your country corresponds to the eighth year of formal schooling, please choose grade 8.

(II) Across grades from preprimary through upper secondary education, at what grade(s) are the topics primarily intended to be taught?

If there are not any specifications to this detail, please indicate national expectations to the best of your ability. If part of a topic does not apply (e.g., energy flow in part A topic (f)), please explain in the comment field.

	(I) Proportion of grade 8 students expected to be taught topic			(II) Grade(s) topic is expected to be taught preprimary (PP) through the end of upper secondary (G12)											
	Check one circle for each line.			Check the corresponding grade(s) for each topic											
B. Chemistry	All or almost all students	Only the more able students	Not Included in the curriculum through grade 8	PP	G1	G2	G3	G4	G5	G6	G7	G8	G9	G10	G11	G12
a) Classification, composition, and particulate structure of matter (elements, compounds, mixtures, molecules, atoms, protons, neutrons, electrons)	○	○	○	☐	☐	☐	☐	☐	☐	☐	☐	☐	☐	☐	☐	☐
b) Physical and chemical properties of matter	○	○	○	☐	☐	☐	☐	☐	☐	☐	☐	☐	☐	☐	☐	☐
c) Mixtures and solutions (solvent, solute, concentration/dilution, effect of temperature on solubility)	○	○	○	☐	☐	☐	☐	☐	☐	☐	☐	☐	☐	☐	☐	☐
d) Properties and uses of common acids and bases	○	○	○	☐	☐	☐	☐	☐	☐	☐	☐	☐	☐	☐	☐	☐
e) Chemical change (transformation of reactants, evidence of chemical change, conservation of matter, common oxidation reactions – combustion, rusting, tarnishing)	○	○	○	☐	☐	☐	☐	☐	☐	☐	☐	☐	☐	☐	☐	☐
f) The role of electrons in chemical bonds	○	○	○	☐	☐	☐	☐	☐	☐	☐	☐	☐	☐	☐	☐	☐

Comments:

References

Achieve. (2004). *Do graduation tests measure up? A closer look at state high school exit exams.* Retrieved from Washington, DC: www.achieve.org/files/TestGraduation-FinalReport.pdf

Adams, R. J., Wu, M. L., & Macaskill, G. (1997). Scaling methodology and procedures for the mathematics and science scales. In M. O. Martin & D. L. Kelly (Eds.), *Third International Mathematics and Science Study: Technical report* (Vol. II: Implementation and Analysis, pp. 111–146). Boston, MA: Center for the Study of Testing, Evaluation, and Educational Policy, Boston College.

Airasian, P. W., & Madaus, G. F. (1983). Linking testing and instruction: Policy issues. *Journal of Educational Measurement, 20*(2), 103–118.

Anderson, C. A. (1967). Construction of occupational classification scheme. In T. Husén (Ed.), *International Study of Achievement in Mathematics: A Comparison of Twelve Countries* (Vol. I, pp. 139–146). New York, NY: John Wiley & Sons.

Averch, H. A., Carroll, S. J., Donaldson, T. S., Kiesling, H. J., & Pincus, J. (1974). *How Effective Is Schooling? A Critical Review of Research.* Englewood Cliffs, NJ: Educational Technology Publications.

Baker, E. L., Chung, G. K. W. K., & Cai, L. (2016). Assessment gaze, refraction, and blur: The course of achievement testing in the past 100 years. *Review of Research in Education, 40*(1), 94–142. doi: 10.3102/0091732X16679806.

Baker, D. P., Goesling, B., & LeTendre, G. K. (2002). Socioeconomic status, school quality, and national economic development: A cross-national analysis of the "Heyneman-Loxley Effect" on mathematics and science achievement. *Comparative Education Review, 46,* 291–312. doi:10.1086/341159.

Beaton, A. E., Martin, M. O., Mullis, I., Gonzalez, E. J., Smith, T. A., & Kelly, D. L. (1996a). *Science Achievement in the Middle School Years: IEA's Third International Mathematics and Science Study.* Chestnut Hill, MA: Center for the Study of Testing, Evaluation, and Educational Policy, Boston College.

Beaton, A. E., Mullis, I., Martin, M. O., Gonzalez, E. J., Kelly, D. L., & Smith, T. A. (1996b). *Mathematics Achievement in the Middle School Years: IEA's Third International Mathematics and Science Study.* Chestnut Hill, MA: Center for the Study of Testing, Evaluation, and Educational Policy, Boston College.

Berliner, D. C. (1990). What's all the fuss about instructional time? In M. Ben-Perez & R. Bromme (Eds.), *The Nature of Time in School* (pp. 3–35). New York, NY: Teachers College Press.

Bidwell, C. E., & Kasarda, J. D. (1980, August). Conceptualising and measuring the effects of school and schooling. *American Journal of Education, 88*(4), 401–430.

Bloom, B. S. (1968). Learning for mastery. Instruction and curriculum. Regional Education Laboratory for the Carolinas and Virginia, Topical Papers and Reprints, Number 1. Evaluation Comment, 1(2). Retrieved from http://eric.ed.gov/?id=ED053419

(1974). Time and learning. *American Psychologist, 29*(9), 682–688.

(1976). *Human Characteristics and School Learning*. New York, NY: McGraw-Hill.

Bloom, B. S., Bjorkquist, L. M., Foshay, A. W., Groen, M., Harada, T., Husén, T., ... Wiegersma, S. (1967). Social factors in education. In T. Husén (Ed.), *International Study of Achievement in Mathematics: A Comparison of Twelve Countries* (Vol. II, pp. 199–259). New York, NY: John Wiley & Sons.

Borg, W. R. (1980). Time and school learning. In C. Denham & A. Lieberman (Eds.), *Time to Learn: A Review of the Beginning Teacher Evaluation Study* (pp. 33–72). Washington, DC: National Institute of Education.

Bowles, S., & Levin, H. M. (1968). The determinants of scholastic achievement – An appraisal of some recent evidence. *The Journal of Human Resources, 3*(1), 324.

Bradburn, N., Haertel, E., Schwille, J., & Torney-Purta, J. (1991). A rejoinder to "I Never Promised You First Place" on JSTOR. *Phi Delta Kappan*, 774–777. Retrieved from www.jstor.org/stable/20404533?seq=1#page_scan_tab_contents

Bradley, R. H., & Corwyn, R. F. (2002). Socioeconomic status and child development. *Annual Review of Psychology, 53*, 371–399.

Broadfoot, P. (2010). Comparative education for the 21st century: Retrospect and prospect. *Comparative Education, 36*(3), 357–371. doi: 10.1080/03050060050129036.

Brown, J., Schiller, K., Roey, S., Perkins, R., Schmidt, W. H., & Houang, R. (2013). *Algebra I and Geometry Curricula* (NCES 2013–451). Washington, DC: U.S. Government Printing Office.

Buchmann, C. (2002). Measuring family background in international studies of education: Conceptual issues and methodological challenges. In A. C. Porter & A. Gamoran (Eds.), *Methodological Advances in Cross-National Surveys of Educational Achievement* (pp. 150–197). Washington, DC: National Academies Press.

Burstein, L. (1991). [Project internal working document presented at TIMSS-95 country planning meeting]. Unpublished internal working document.

(Ed.). (1993). *The IEA Study of Mathematics III: Student Growth and Classroom Processes* (Vol. 3). Oxford: Pergamon Press.

Carnoy, M., Khavenson, T., & Ivanova, A. (2015). Using TIMSS and PISA results to inform educational policy: A study of Russia and its neighbours.

Compare: A Journal of Comparative and International Education, *45*(2), 248–271. doi:10.1080/03057925.2013.855002.

Carnoy, M., Khavenson, T., Loyalka, P., Schmidt, W. H., & Zakharov, A. (2016). Revisiting the relationship between international assessment outcomes and educational production: Evidence from a longitudinal PISA-TIMSS sample. *American Educational Research Journal*, *53*(4), 1054–1085. doi: 10.3102/0002831216653180.

Carnoy, M., & Rothstein, R. (2013). What do international tests really show about US student performance. *Economic Policy Institute*, *28*.

Carroll, J. B. (1963). A model of school learning. *Teachers College Record*, *64*(8), 723–733.

(1975). *The Teaching of French as a Foreign Language in Eight Countries*. New York, NY: Wiley.

(1984). The model of school learning: Progress of an idea. In L. W. Anderson (Ed.), *Time and School Learning: Theory, Research and Practice* (pp. 15–45). London: Cambridge University Press Archive.

(1989). The Carroll Model: A 25-year retrospective and prospective view. *Educational Researcher*, *18*(1), 26–31. doi:10.3102/0013189X018001026.

Causa, O., & Chapuis, C. (2009). *Equity in Student Achievement across OECD Countries: An Investigation of the Role of Policies*. (OECD Economics Department Working Papers, No. 708). doi: 10.1787/223056645650.

Chapin, F. S. (1928). A quantitative scale for rating the home and social environment of middle class families in an urban community: A first approximation to the measurement of socio-economic status. *Journal of Educational Psychology*, *19*(2), 99–111. doi:10.1037/h0074500.

Chudgar, A., & Luschei, T. F. (2009). National income, income inequality, and the importance of schools: A hierarchical cross-national comparison. *American Educational Research Journal*, *46*(3), 626–658.

Clements, M. A. K., Bishop, A. J., Keitel, C., Kilpatrick, J., & Leung, F. K. S. (2013). From the few to the many: Historical perspectives on who should learn mathematics. In M. A. K. Clements, A. J. Bishop, C. Keitel, J. Kilpatrick, & F. K. S. Leung (Eds.), *Third International Handbook of Mathematics Education* (pp. 7–40). New York, NY: Springer.

Cogan, L. S., & Schmidt, W. H. (2015). The concept of opportunity to learn (OTL) in international comparisons of education. In K. Stacey & R. Turner (Eds.), *Assessing Mathematical Literacy: The PISA Experience* (pp. 207–216). Switzerland: Springer International Publishing. doi: 10.1007/978-3-319-10121-7.

Cogan, L. S., Schmidt, W. H., & Houang, R. (2014). Primary teacher preparation in the United States: What we have learned. In S. Blömeke, F.-J. Hsieh, G. Kaiser, & W. H. Schmidt (Eds.), *International Perspectives on Teacher Knowledge, Beliefs, and Opportunities to Learn* (pp. 355–369). Dordrecht/Heidelberg/New York/London: Springer.

Cogan, L. S., Schmidt, W. H., & Wiley, D. E. (2001). Who takes what math and in which track? Using TIMSS to characterize U.S. students' eighth-grade mathematics learning opportunities. *Educational Evaluation and Policy Analysis*, *23*(4), 323–341.

Coleman, J. S. (1975). Methods and results in the IEA studies of effects of school on learning. *Review of Educational Research*, *45*(3), 335–386.

Coleman, J. S., Campbell, E. Q., Hobson, C. J., McPartland, J., Mood, A. M., Weinfeld, F. D., & York, R. L. (1966). *Equality of educational opportunity*, (NCES No. OE-38001). Retrieved from https://files.eric.ed.gov/fulltext/ED012275.pdf.

Common Core State Standards Initiative. (2010). Common Core State Standards for Mathematics (p. 93): National Governors Association Center for Best Practices (NGA Center) and the Council of Chief State School Officers (CCSSO).

Cowen, R. (1996). Last past the post: Comparative education, modernity and perhaps post-modernity. *Comparative Education*, *32*(2), 151–170. doi: 10.1080/03050069628812.

Darling-Hammond, L., & Snyder, J. (1992). Curriculum studies and the traditions of inquiry: the scientific tradition. In P. W. Jackson (Ed.), *Handbook of Research on Curriculum: A Project of the American Educational Research Association*. New York, NY: Simon & Schuster Macmillan.

de Lange, J. (1996). Using and applying mathematics in education. In A. J. Bishop, K. Clements, C. Keitel, J. Kilpatrick, & C. Laborde (Eds.), *International Handbook of Mathematics Education* (Part 1, pp. 49–97). Dordrecht: Kluwer.

(2003). Mathematics for literacy. In B. L. Madison & L. A. Steen (Eds.), *Quantitative Literacy: Why Numeracy Matters for Schools and Colleges* (pp. 75–89). Princeton, NJ: The National Council on Education and the Disciplines.

Dumay, X., & Dupriez, V. (2007). Accounting for class effect using the TIMSS 2003 eighth-grade database: Net effect of group composition, net effect of class process, and joint effect. *School Effectiveness and School Improvement*, *18*(4), 383–408. doi:10.1080/09243450601146371.

Elliott, J., & Kushner, S. (2007). The need for a manifesto for educational programme evaluation. *Cambridge Journal of Education*, *37*(3), 321–336. doi: 10.1080/03057640701546649.

Entwisle, D. R., & Astone, N. M. (1994). Some practical guidelines for measuring youth's race/ethnicity and socioeconomic status. *Child Development*, *65*(6), 1521–1540. doi: 10.2307/1131278.

Foshay, A. W. (1962). The background and the procedures of the twelve-country study: Educational achievements of thirteen-year-olds in twelve countries (pp. 7–19). Hamburg: UNESCO Institute for Education.

Foshay, A. W., Thorndike, R. L., Hotyat, F., Pidgeon, D. A., & Walker, D. A. (1962). *Educational Achievements of Thirteen-Year-Olds in Twelve Countries*. Hamburg. Retrieved from http://unesdoc.unesco.org/images/0013/001314/131437eo.pdf.

Floden, R. E. (2002). The Measurement of Opportunity to Learn. In A. C. Porter & A. Gamoran (Eds.), *Methodological Advances in Cross-National Surveys of*

Educational Achievement (pp. 231–266). Washington, DC: National Academies Press.

Fuller, B., & Clarke, P. (1994). Raising school effects while ignoring culture? Local conditions and the influence of classroom tools, rules, and pedagogy. *Review of Educational Research, 64*(1), 119–157.

Gamoran, A. (1996). Effects of schooling on children and families. In A. Booth and J. F. Dunn (Eds.), *Family-School Links: How Do They Affect Education Outcomes?* (pp. 107–114) Hillsdale, NJ: Erlbaum.

Gamoran, A., & Berends, M. (1987). The effects of stratification in secondary schools: Synthesis of survey and ethnographic research. *Review of Educational Research, 57*(4), 415–435.

Gamoran, A., & Long, D. A. (2007). Equality of educational opportunity: A 40 year retrospective. In *International Studies in Educational Inequality, Theory and Policy* (pp. 23–47). Dordrecht: Springer Netherlands. doi: 10.1007/978-1-4020-5916-2_2.

Ganzeboom, H. B. G., De Graaf, P. M., & Treiman, D. J. (1992). A standard international socio-economic index of occupational status. *Social Science Research, 21*(1), 1–56. doi: 10.1016/0049-089X(92)90017-B.

Garden, R. A., & Livingstone, I. (1989). The contexts of mathematics education: Nations, communities and schools. In D. F. Robitaille & R. A. Garden (Eds.), *The IEA Study of Mathematics II: Contexts and Outcomes of School Mathematics* (pp. 17–38). Oxford: Pergamon Press.

Garden, R. A., & Orpwood, G. (1996). Development of the TIMSS achievement tests. In M. O. Martin & D. L. Kelly (Eds.), *Third International Mathematics and Science Study Technical Report* (Vol. Volume I: Design and Development, pp. 2-1-2-19). Chestnut Hill, MA: Boston College.

Goldin, G. (2008). Perspectives on representation in mathematical learning and problem solving. In L. D. English, M. Bartolini Bussi, G. A. Jones, R. A. Lesh, B. Sriraman, & D. Tirosh (Eds.), *Handbook of international research in mathematics education* (pp. 176–201). New York, NY: Routledge.

Gonzales, P., Calsyn, C., Jocelyn, L., Mak, K., Kastberg, D., Arafeh, S., . . . Tsen, W. (2000). *Pursuing Excellence: Comparisons of International Eighth-Grade Mathematics and Science Achievement from a U.S. Perspective, 1995 and 1999: Initial Findings from the Third International Mathematics and Science Study-Repeat*. Washington, DC: DIANE Publishing.

Greenwald, R., Hedges, L. V, & Laine, R. D. (1996). The effect of school resources on student achievement. *Review of Educational Research, 66*(3), 361–396.

Grek, S. (2009). Governing by numbers: the PISA "effect" in Europe. *Journal of Education Policy, 24*(1), 23–37. doi: 10.1080/02680930802412669.

Gustafsson, J.-E. (2008). Effects of international comparative studies on educational quality on the quality of educational research. *European Educational Research Journal, 7*(1), 1. doi: 10.2304/eerj.2008.7.1.1.

Gustafsson, J.-E., & Rosen, M. (2014). Quality and credibility of international studies. In R. Strietholt, W. Bos, J.-E. Gustafsson, & M. Rosen (Eds.),

Educational Policy Evaluation through International Comparative Assessments (pp. 19–31). Münster: Waxmann Verlag.

Hans, N. (1949). *Comparative Education: A Study of Educational Factors and Traditions*. London: Routledge & Kegan Paul Limited.

Hanushek, E. A., & Kain, J. F. (1972). On the value of equality of educational opportunity as a guide to public policy. In F. Mosteller & D. P. Moynihan (Eds.), *On Equality of Educational Opportunity* (pp. 116–145). New York, NY: Vintage Books.

Hanushek, E. A., & Woessmann, L. (2007). *The role of education quality for economic growth* (World Bank Policy Research Working Paper 4122). Retrieved from http://documents.worldbank.org/curated/en/260461468324885735/pdf/wps4122.pdf.

Harlen, W. (2001). The assessment of scientific literacy in the OECD/PISA project. *Studies in Science Education, 36*(1), 79–103. doi: 10.1080/03057260108560168.

Harmon, M., Smith, T. A., Martin, M. O., Kelly, D. L., Beaton, A. E., Mullis, I., . . . Orpwood, G. (1997). *Performance Assessment in IEA's Third International Mathematics and Science Study*. Chestnut Hill, MA: Center for the Study of Testing, Evaluation, and Educational Policy, Boston College.

Hauser, R. M. (1994). Measuring socioeconomic status in studies of child development. *Child Development, 65*(6), 1541–1545.

Hauser, R. M., & Featherman, D. L. (1976). Equality of schooling: Trends and prospects. *Sociology of Education, 29*(2), 99–120.

Heyneman, S. P., & Loxley, W. A. (1983). The effect of primary-school quality on academic achievement across twenty-nine high- and low-income countries. *American Journal of Sociology, 88*(6), 1162–1194.

Hiebert, J., & Stigler, J. W. (2000). A proposal for improving classroom teaching: Lessons from the TIMSS Video Study. *The Elementary School Journal, 101*(1), 3–20.

Holliday, W. G., & Holliday, B. W. (2003). Why using international comparative math and science achievement data from TIMSS is not helpful. *The Educational Forum, 67*(3), 250–257. doi: 10.1080/00131720309335038.

Houang, R. T., & Schmidt, W. H. (2008). TIMSS international curriculum analysis and measuring educational opportunities. In *3rd IEA International Research Conference, Taipei, Chinese Taipei* (p. 18). Taipei, Chinese Taipei.

Howie, S. J., & Plomp, T. (2006). Lessons from cross-national research on context and achievement: Hunting and fishing in the TIMSS landscape. In S. J. Howie & T. Plomp (Eds.), *Contexts of Learning Mathematics and Science: Lessons Learned from TIMSS* (pp. 3–16). Abingdon, Oxon: Routledge.

Husén, T. (Ed.). (1967a). *International Study of Achievement in Mathematics: A Comparison of Twelve Countries* (Vol. I). New York, NY: Wiley.

(Ed.). (1967b). *International Study of Achievement in Mathematics: A Comparison of Twelve Countries* (Vol. II). New York, NY: Wiley.

(1974). Multi-national evaluation of school systems: Purposes, methodology, and some preliminary findings. *Scandinavian Journal of Educational Research, 18*(1), 13–39.

(1979). An international research venture in retrospect: The IEA surveys. *Comparative Education Review, 23*(3), 371–385. doi: 10.1086/446067.

(1983). Are standards in U.S. schools really lagging behind those in other countries? *The Phi Delta Kappan, 64*(7), 455–461.

Husén, T., & Postlethwaite, N. (1996). A brief history of the International Association for the Evaluation of Educational Achievement (IEA). *Assessment in Education: Principles, Policy & Practice, 3*(2), 129–141. doi: 10.1080/0969594960030202.

IEA. Home. (n.d.). Retrieved from www.iea.nl/.

International Association for the Evaluation of Educational Achievement (IEA). (1988). *Science Achievement in Seventeen Countries: A Preliminary Report.* Oxford: Pergamon.

Institute of Education Sciences. (n.d.). *The middle grades longitudinal study of 2017–18 (MGLS: 2017).* Retrieved from https://nces.ed.gov/surveys/mgls/.

Jakwerth, P. M. (1996). Evaluating content validity in cross-national achievement tests. (Ph.D. Dissertation), Michigan State University, East Lansing.

James, W. (1899). *Talks to Teachers on Psychology: And to Students on Some of Life's Ideals.* New York, NY: H. Holt and company.

Jencks, C., Smith, M., Acland, H., Bane, M. S., Cohen, D., Gintis, H., Heyns, B., & Michelson, S. (1972). *Inequality: A Reassessment of the Effect of Family and Schooling in America.* New York, NY: Basic Books.

Keeves, J. P. (1974). The IEA science project: Science achievement in three countries – Australia, the Federal Republic of Germany and the United States. In M. Bruderlin (Ed.), *Implementation of Curricula in Science Education* (pp. 158–178). Cologne: German Commission for UNESCO.

(1992). *Learning Science in a Changing World. Cross-National Studies of Science Achievement: 1970 to 1984.* Hague, The Netherlands: IEA International Headquarters.

(2011). IEA - from the beginning in 1958 to 1990. In C. Papanastasiou, T. Plomp, & E. C. Papanastasiou (Eds.), *IEA 1958–2008: 50 Years of Experiences and Memories* (Vol. 1). Amsterdam: International Association for the Evaluation of Educational Achievement.

Keeves, J. P., & Lietz, P. (2011). The relationship of IEA to some developments in educational research methodology and measurement during the years from 1962 to 1992. C. Papanastasiou, T. Plomp, & E. C. Papanastasiou (Eds.), *IEA 1958–2008: 50 Years of Experiences and Memories* (Vol. 1, pp. 217–251). Amsterdam: International Association for the Evaluation of Educational Achievement.

Krathwohl, D. R., Bloom, B. S., & Masia, B. B. (1956). *Taxonomy of Educational Objectives: The Classification of Educational Goals.* New York, NY: David McKay.

Lucas, S. R. (1999). *Tracking Inequality: Stratification and Mobility in American High Schools* (Sociology of Education Series). New York, NY: Teachers College Press.

Lundgren, U. P. (2011). PISA as a political instrument. In M. A. Pereyra, H.-G. Kotthoff, & R. Cowen (Eds.), *PISA under Examination: Changing Knowledge, Changing Tests, and Changing Schools* (pp. 17–30). Rotterdam, NY: SensePublishers.

Martin, M. O., & Kelly, D. L. (Eds.). (1996). *Third International Mathematics and Science Study: Technical report (volume 1): Design and development.* Chestnut Hills, MA: Center for the Study of Testing, Evaluation, and Educational Policy, Boston College.

Martin, M. O., & Mullis, I. V. S. (2006). TIMSS: Purpose and design. In S. J. Howie & T. Plomp (Eds.), *Contexts of Learning Mathematics and Science: Lessons Learned from TIMSS* (pp. 17–30). Abingdon, Oxon: Routledge.

Martin, M. O., Mullis, I., Beaton, A. E., Gonzalez, E. J., Kelly, D. L., & Smith, T. A. (1997). *Science Achievement in the Primary School Years: IEA's Third International Mathematics and Science Study.* Chestnut Hill, MA: Center for the Study of Testing, Evaluation, and Educational Policy, Boston College.

McDonnell, L. M. (1995). Opportunity to learn as a research concept and a policy instrument. *Educational Evaluation and Policy Analysis, 17*(3), 305–322. doi: 10.3102/01623737017003305.

Medrich, E. A., & Griffith, J. E. (1992). *International Mathematics and Science Assessment: What Have We Learned?* (NCES 92–011). Washington, DC: US Department of Education. Retrieved from http://nces.ed.gov/pubs92/92011.pdf.

Michalowicz, K. D., & Howard, A. C. (2003). Pedagogy in text: An analysis of mathematics texts from the nineteenth century. In G. M. A. Stanic & J. Kilpatrick (Eds.), *A History of School Mathematics* (Vol. 1, pp. 77–109). Reston, VA: National Council of Teachers of Mathematics.

Mislevy, R. J. (1991). Randomization-based inference about latent variables from complex samples. *Psychometrika, 56*(2), 177–196.

Mitter, W. (1997). Challenges to comparative education: Between retrospect and expectation. *International Review of Education, 43*(5–6), 401–412. doi: 10.1023/A:1003084402042.

Montt, G. (2011). Cross-national differences in educational achievement inequality. *Sociology of Education, 84*(1), 49–68.

Morgan, C., & Shahjahan, R. A. (2014). The legitimation of OECD's global educational governance: Examining PISA and AHELO test production. *Comparative Education, 50*(2), 192–205.

Mullis, I., Martin, M. O., Beaton, A. E., Gonzalez, E. J., Kelly, D. L., & Smith, T. A. (1997). *Mathematics Achievement in the Primary School Years: IEA's Third International Mathematics and Science Study.* Chestnut Hill, MA: Center for the Study of Testing, Evaluation, and Educational Policy, Boston College.

Mullis, I. V. S., Martin, M. O., & Foy, P. (2008). *TIMSS 2007 International Mathematics Report: Findings from IEA's Trends in International Mathematics and Science Study at the Fourth and Eighth Grades.* Chestnut Hill, MA: TIMSS & PIRLS International Study Center, Lynch School of Education, Boston College.

Mullis, I. V. S., Martin, M. O., Foy, P., & Hooper, M. (2016). *TIMSS 2015 International Results in Mathematics.* Chestnut Hill, MA: TIMSS & PIRLS International Study Center, Lynch School of Education, Boston College.

Mundy, K., & Ghali, M. (2009). International and transnational policy actors in education: A review of the research. In G. Sykes, B. Schneider, & D. Plank (Eds.), *Handbook of Education Policy Research* (pp. 717–734). New York, NY: Routledge.

National Center for Education Statistics. (2012). *Improving the Measurement of Socioeconomic Status for the National Assessment of Educational Progress: A Theoretical Foundation.* Retrieved from Washington, DC: http://nces.ed.gov/nationsreportcard/pdf/researchcenter/Socioeconomic_Factors.pdf.

National Governors Association Center for Best Practices & Council of Chief State School Officers (2010). *Common Core State Standards for Mathematics.* Retrieved from www.corestandards.org/Math/.

National Mathematics Advisory Panel. (2008). *Foundations for Success: The Final Report of the National Mathematics Advisory Panel.* Washington, DC: US Department of Education.

National Research Council. (2012). *Education for Life and Work: Developing Transferable Knowledge and Skills in the 21st Century.* Washington, DC: The National Academies Press. doi: 10.17226/13398.

Noah, H. J., & Eckstein, M. A. (1969). *Toward a Science of Comparative Education.* New York, NY: The Macmillan Company.

Nonoyama-Tarumi, Y., & Willms, J. D. (2010). The relative and absolute risks of disadvantaged family background and low levels of school resources on student literacy. *Economics of Education Review, 29*(2), 214–224. doi: 10.1016/j.econedurev.2009.07.007.

Oakes, J. (1985). *Keeping Track: How Schools Structure Inequality.* New Haven, CT: Yale University Press.

OECD. (2000). *Measuring Student Knowledge and Skills: The PISA 2000 Assessment of Reading, Mathematical and Scientific Literacy.* Paris: OECD Publishing.

(2004). *The PISA 2003 Assessment Framework: Mathematics, Reading, Science and Problem Solving Knowledge and Skills.* Paris: OECD Publishing.

(2013a). *PISA 2012 Assessment and Analytical Framework: Mathematics, Reading, Science, Problem Solving and Financial Literacy.* Paris: OECD Publishing. doi: 10.1787/9789264190511-en.

(2013b). *PISA 2012 Results: Excellence through Equity: Giving Every Student the Chance to Succeed (Volume II).* Paris: OECD Publishing.

(2014a). *PISA 2012 Results: What Students Know and Can Do (Volume I, Revised Edition, February 2014): Student Performance in Mathematics, Reading and Science.* Paris: OECD Publishing.

(2014b). *PISA 2012 technical report*. Retrieved from www.oecd.org/pisa/pisa products/PISA-2012-technical-report-final.pdf.

(2015). *PISA 2015 Draft Science Framework*. OECD Publishing. Retrieved from www.oecd.org/pisa/pisaproducts/Draft%20PISA%202015%20Science%20Framework%20.pdf.

(n.d.). About PISA, retrieved September 30, 2015, from www.oecd.org/pisa/aboutpisa/

Pant, H. A., Tiffin-Richards, S. P., & Stanat, P. (2017). Standards setting: Bridging the worlds of policy making and research. In S. Blömeke & J. -E. Gustafsson (Eds.), *Standard Setting in Education. Methodology of Educational Measurement and Assessment* (pp. 49–68). New York, NY: Springer.

Pereyra, M. A., Kotthoff, H. -G., & Cowen, R. (2011). PISA under examination. In M. A. Pereyra, H.-G. Kotthoff, & R. Cowen (Eds.), *PISA under Examination: Changing Knowledge, Changing Tests, and Changing Schools* (pp. 1–14). Rotterdam: SensePublishers. doi: 10.1007/978-94-6091-740-0.

PISA 2021 Mathematics Advisory Committee. (2017). *Mathematical Literacy: A Broadened Perspective*. (Working Paper Submitted to PISA Governing Board). Paris: OECD.

Plowden, B. (Ed.). (1967a). *Children and their Primary Schools: A Report of the Control Advisory Council for Education (England)* (Volume I: The Report). London: Her Majesty's Stationery Office.

(Ed.). (1967b). *Children and their Primary Schools: A Report of the Control Advisory Council for Education (England)* (Volume II: Research and Surveys). London: Her Majesty's Stationery Office. Retrieved from www.educationengland.org.uk/documents/plowden/plowden1967-2.html.

Postlethwaite, T. N. (1967). *School Organization and Student Achievement: A Study Based on Achievement in Mathematics in Twelve Countries*. Stockholm: Almqvist & Wiksell.

Postlethwaite, T. N., & Wiley, D. E. (Eds.). (1992). *The IEA Study of Science I: Science Achievement in Twenty-Three Countries*. Oxford: Pergamon Press.

Purves, A. C. (1987). The Evolution of the IEA: A Memoir. *Comparative Education Review*, *31*(1), 10–28.

Reimers, F. M., & Chung, C. K. (2016). *Teaching and Learning for the Twenty-First Century*. Cambridge, MA: Harvard Education Press.

Riddell, A. R. (1989). An alternative approach to the study of school effectiveness in third world countries. *Comparative Education Review*, *33*(4), 481–497.

Robitaille, D. F. (1989). Students' achievements: Population A. In D. F. Robitaille & R. A. Garden (Eds.), *The IEA Study of Mathematics II: Contexts and Outcomes of School Mathematics* (pp. 102–125). Oxford: Pergamon Press.

Robitaille, D. F., & Garden, R. A. (Eds.). (1989). *The IEA Study of Mathematics II: Contexts and Outcomes of School Mathematics* (Vol. 2). Oxford: Pergamon Press.

Robitaille, D. F., Schmidt, W. H., Raizen, S. A., McKnight, C. C., Britton, E., & Nicol, C. (1993). *Curriculum Frameworks for Mathematics and Science*.

TIMSS Monograph No. 1. Vancouver: Pacific Educational Press, Faculty of Education, University of British Columbia.

Romberg, T. A., & Zarinnia, A. (1987). Consequences of the new world view to assessment of students' knowledge of mathematics. In T. A. Romberg & D. M, Stewart (Eds.), *The Monitoring of School Mathematics: Background Papers* (Vol. 2, pp. 153–201). Madison, WI: Wisconsin Center for Education Research, School of Education, University of Wisconsin-Madison.

Rosier, M. J., & Keeves, J. P. (Eds.). (1991). *The IEA Study of Science I: Science Education and Curricula in Twenty-Three Countries*. Oxford: Pergamon Press.

Samoff, J. (2012). Institutionalizing international influence. In R. F. Arnove & C. A. Torres (Eds.), *Comparative Education: The Dialectic of the Global and the Local* (pp. 55–85). Lanham, MD: Rowman & Littlefield Publishers, Inc.

Schiller, K. S., Schmidt, W. H., Muller, C., & Houang, R. T. (2010). Hidden disparities: How courses and curricula shape opportunities in mathematics during high school. *Equity & Excellence in Education*, *43*(4), 414–433. doi: 10.1080/10665684.2010.517062.

Schmidt, W. H. (2009). *Exploring the Relationship between Content Coverage and Achievement: Unpacking the Meaning of Tracking in Eighth-Grade Mathematics*. Education Policy Center, East Lansing, MI: Michigan State University.

(2017, Winter). Excellence and equality in mathematics education. *American Affairs*, *1*(4): 32–33.

Schmidt, W. H., & Burroughs, N. A. (2016). The trade-off between excellence and equality: What international assessments tell us. *Georgetown Journal of International Affairs*, *17*(1), 103–109. doi: 10.1353/gia.2016.0011.

Schmidt, W. H., Burroughs, N. A., Cogan, L. S., & Houang, R. T. (2017). The role of subject-matter content in teacher preparation: an international perspective for mathematics. *Journal of Curriculum Studies*, *49*(2), 111–131. doi: 10.1080/00220272.2016.1153153.

Schmidt, W. H., Burroughs, N. A., Zoido, P., & Houang, R. T. (2015). The role of schooling in perpetuating educational inequality: An international perspective. *Educational Researcher*, *44*(7), 371–386. doi:10.3102/0013189X15603982.

Schmidt, W. H., & Cogan, L. S. (1996). Development of the TIMSS context questionnaires. In M. O. Martin & D. L. Kelly (Eds.), *Third International Mathematics and Science Study: Technical Report* (Vol. Volume I: Design and Development, pp. 5-1–5-22). Chestnut Hill, MA: Boston College.

(2009). The myth of equal content. *Educational Leadership*, *67*(3), 44–47.

(2014). Greater expectations in lower secondary mathematics teacher preparation: An examination of future teachers' opportunity to learn profiles. In S. Blömeke, F.-J. Hsieh, G. Kaiser, & W. H. Schmidt (Eds.), *International Perspectives on Teacher Knowledge, Beliefs, and Opportunities to Learn* (pp. 393–414). Dordrecht/ Heidelberg/New York/London: Springer.

Schmidt, W. H., Cogan, L. S., & Guo, S. (in press). The role that mathematics plays in college- and career-readiness: Evidence from PISA. *Journal of Curriculum Studies*.

Schmidt, W. H., Cogan, L., & Houang, R. (2011). The role of opportunity to learn in teacher preparation: An international context. *Journal of Teacher Education*, *62*(2), 138–153.

(2014). Emphasis and balance among the components of teacher preparation: The case of lower-secondary mathematics teacher education. In S. Blömeke, F.-J. Hsieh, G. Kaiser, & W. H. Schmidt (Eds.), *International Perspectives on Teacher Knowledge, Beliefs, and Opportunities to Learn* (pp. 371–392). Dordrecht/Heidelberg/New York/London: Springer.

Schmidt, W. H., Cogan, L. S., Houang, R. T., & McKnight, C. C. (2011). Content coverage differences across districts/states: A persisting challenge for U.S. education policy. *American Journal of Education*, *117*(3), 399–427.

Schmidt, W. H., Cogan, L. S., & McKnight, C. C. (2010). Equality of educational opportunity: myth or reality in U.S. schooling? *American Educator*, *34*(4), 12–19.

Schneider, B., Carnoy, M., Kilpatrick, J., Schmidt, W. H., & Shavelson, R. J. (2007). *Estimating Causal Effects Using Experimental and Observational Designs* (report from the Governing Board of the American Educational Research Association Grants Program). Washington, DC: American Educational Research Association.

Schmidt, W. H., & Houang, R. T. (2007). Lack of focus in the mathematics curriculum: Symptom or cause? In T. Loveless (Ed.), *Lessons Learned: What International Assessments Tell Us about Math Achievement* (pp. 65 –84). Washington, DC: Brookings Institution Press.

(2012). Curricular coherence and the Common Core State Standards for Mathematics. *Education Researcher*, *41*(8), 294–308.

Schmidt, W., Houang, R., & Cogan, L. (2002). A coherent curriculum: The case of mathematics. *American Educator*, *26*(2), 1–17.

(2011). Preparing future math teachers. *Science*, *332*(603), 1266–1267.

Schmidt, W. H., Houang, R., & Shakrani, S. (2009). *International Lessons about National Standards*. Washington, DC: The Thomas B. Fordham Institute.

Schmidt, W. H., Jakwerth, P. M., & McKnight, C. C. (1998). Curriculum-sensitive assessment: Content *does* make a difference. *International Journal of Educational Research*, *29*(6), 503–527.

Schmidt, W. H., Jorde, D., Cogan, L. S., Barrier, E., Gonzalo, I., Moser, U., ... Wolfe, R. G. (1996). *Characterizing Pedagogical Flow: An Investigation of Mathematics and Science Teaching in Six Countries*. Dordrecht/Boston/London: Kluwer.

Schmidt, W. H., & McKnight, C. C. (1995). Surveying educational opportunity in Mathematics and Science: An International Perspective. *Educational Evaluation and Policy Analysis*, *17*(3), 337–353. Retrieved from http://epa.sagepub.com/content/17/3/337.full.pdf.

(2012). *Inequality For All: The Challenge of Unequal Opportunity in American Schools*. New York, NY: Teachers College Press.

Schmidt, W. H., McKnight, C., Cogan, L. S., Jakwerth, P. M., & Houang, R. T. (1999). *Facing the Consequences: Using TIMSS for a Closer Look at U.S. Mathematics and Science Education*. Dordrecht/Boston/London: Kluwer.

Schmidt, W. H., McKnight, C. C., Houang, R. T., Wang, H. A., Wiley, D. E., Cogan, L. S., & Wolfe, R. G. (2001). *Why Schools Matter: A Cross-National Comparison of Curriculum and Learning.* San Francisco, CA: Jossey-Bass.
Schmidt, W. H., McKnight, C. C., & Raizen, S. A. (1997). *A splintered Vision: An Investigation of U.S. Science and Mathematics Education.* Dordrecht; Boston: Kluwer Academic Publishers.
Schmidt, W. H., McKnight, C. C., Valverde, G. A., Houang, R. T., & Wiley, D. E. (1997). *Many Visions, Many Aims (Volume I): A Cross-National Investigation of Curricular Intentions in School Mathematics.* Dordrecht, The Netherlands: Kluwer Academic Publishers.
Schmidt, W. H., Raizen, S. A., Britton, E. D., Bianchi, L. J., & Wolfe, R. G. (1997). *Many Visions, Many Aims, (Volume II): A Cross-National Investigation of Curricular Intentions in School Science.* Dordrecht/Boston/London: Kluwer.
Schmidt, W. H., Wang, H. A., & McKnight, C. C. (2005). Curriculum coherence: An examination of US mathematics and science content standards from an international perspective. *Journal of Curriculum Studies, 37*(5), 525–529.
Schmidt, W. H., Zoido, P., & Cogan, L. S. (2014). *Schooling Matters: Opportunity to Learn in PISA 2012* (OECD Education Working Papers, No. 95). Retrieved from doi: 10.1787/5k3vohldmchl-en.
Schwille, J. (2011). Experiencing innovation and capacity building in IEA research, 1963–2008. In C. Papanastasiou, T. Plomp, & E. C. Papanastasiou (Eds.), *IEA 1958–2008: 50 Years of Experiences and Memories* (Vol. II, pp. 627–707). Nicosia: Cultural Center of the Kykkos Monastery.
SciMath[MN]. (2008). *Minnesota TIMSS Report: December 2008 a Preliminary Summary of Results.* St. Paul, MN: SciMath[MN].
Sellar, S., & Lingard, B. (2014). The OECD and the expansion of PISA: new global modes of governance in education. *British Educational Research Journal, 40*(6), 917–936. doi: 10.1002/berj.3120.
Sirin, S. R. (2005). Socioeconomic status and academic achievement: A meta-analytic review of research. *Review of Educational Research, 75*(3), 417–453. doi:10.3102/00346543075003417.
Stevenson, H. W. (1998). A study of three cultures: Germany, Japan, and the United States – An overview of the TIMSS case study project. *Phi Delta Kappan, 79*(7), 524–529.
Stevenson, H. W., & Nerison-Low, R. (1997). *To Sum It Up: Case Studies of Education in Germany, Japan, and the United States.* Washington, DC: National Institute on Student Achievement, Curriculum, and Assessment, Office of Educational Research and Improvement, U.S. Department of Education.
Stigler, J., & Hiebert, J. (1997). Understanding and improving classroom mathematics instruction: An overview of the TIMSS video study. In *Raising Australian Standards in Mathematics and Science: Insights from TIMSS*

Conference Proceedings (pp. 52–65). Camberwell, Victoria: Australian Council for Educational Research.

Stigler, J. W., Gonzales, P., Kawanaka, T., Knoll, S., & Serrano, A. (1999). *The TIMSS Videotape Classroom Study: Methods and Findings from an Exploratory Research Project on Eighth-Grade Mathematics Instruction in Germany, Japan, and the United States*. Washington, DC.

Survey of Mathematics and Science Opportunities. (1993). A description of the TIMSS' achievement test content design: Test blueprints. In *Survey of Mathematics and Science Opportunities Research Report Series*. East Lansing: Michigan State University.

Tatto, M. T., Peck, R., Schwille, J., Bankov, K., Senk, S. L., Rodriguez, M., ... Rowley, G. (2012). *Policy, Practice, and Readiness to Teach Primary and Secondary Mathematics in 17 Countries: Findings from the IEA Teacher Education and Development Study in Mathematics (TEDS-M)*. Amsterdam: International Association for the Evaluation of Educational Achievement (IEA).

Thorndike, R. L. (1962). International comparison of the achievement of 13-year-olds. In A. W. Foshay, R. L. Thorndike, F. Hotyat, D. A. Pidgeon, & D. A. Walker (Eds.), *Educational Achievements of Thirteen-Year Olds in Twelve Countries: Results of an International Research Project, 1959–61* (pp. 21–42). Hamburg: UNESCO Institute for Education.

(1967a). Mathematics test and attitude inventory scores. In T. Husén (Ed.), *International Study of Achievement in Mathematics: A Comparison of Twelve Countries* (Vol. II, pp. 21–48). New York, NY: John Wiley & Sons.

(1967b). The schools, the teachers, and the pupils. In T. Husén (Ed.), *International Study of Achievement in Mathematics: A Comparison of Twelve Countries* (Vol. I, pp. 257–283). New York, NY: John Wiley & Sons.

Torney-Purta, J., & Schwille, J. (1986). Civic values learned in school: Policy and practice in industrialized nations. *Comparative Education Review*, *30*(1), 30–49.

Travers, K. J. (2011). The Second International Mathematics Study (SIMS): Intention, implementation, attainment. In C. Papanastasiou, T. Plomp, & E. C. Papanastasiou (Eds.), *IEA 1958–2008: 50 Years of Experiences and Memories* (Vol. 1, pp. 73–96). Amsterdam: International Association for the Evaluation of Educational Achievement.

Travers, K. J., Oldham, E. E., & Livingstone, I. D. (1989). Origins of the Second International Mathematics Study. In K. J. Travers & I. Westbury (Eds.), *The IEA Study of Mathematics I: Analysis of Mathematics Curricula* (pp. 1–14). Oxford: Pergamon Press.

Travers, K. J., & Westbury, I. (Eds.). (1989). *The IEA Study of Mathematics I: Analysis of Mathematics Curricula* (Vol. 1). Oxford: Pergamon Press.

UNESCO. (1997). *International Standard Classification of Education (ISCED-1997)*. Paris: UNESCO.

(2012). *International Standard Classification of Education (ISCED-2011)*. Montreal, Quebec: UNESCO Institute for Statistics.

US Department of Education. (1996). National Center for Education Statistics, *Pursuing Excellence: A Study of U.S. Twelfth-Grade Mathematics and Science Achievement in International Context* (NCES 97–198). Washington, DC: US Government Printing Office.

Valverde, G. A., Bianchi, L. J., Wolfe, R. G., Schmidt, W. H., & Houang, R. T. (2002). *According to the Book: Using TIMSS to Investigate the Transition from Policy to Pedagogy in the World of Textbooks*. Dordrecht/Boston/London: Kluwer Academic Publishers.

Valverde, G. A., & Schmidt, W. H. (2000). Greater expectations: learning from other nations in the quest for 'world-class standards' in US school mathematics and science. *Journal of Curriculum Studies, 32*(5), 651-687.

Wagemaker, H., & Knight, G. (1986). The students of mathematics. In D. F. Robitaille & R. A. Garden (Eds.), *The IEA Study of Mathematics II: Contexts and Outcomes of School Mathematics* (pp. 63–83). Oxford: Pergamon Press.

Walker, D. A. (1976). *The IEA Six Subject Survey: An Empirical Study of Education in Twenty-One Countries*. New York, NY: Halsted Press.

(1962). An analysis of the reactions of Scottish teachers and pupils to items in the geography, mathematics and science tests. In A. W. Foshay, R. L. Thorndike, F. Hotyat, D. A. Pidgeon, & D. A. Walker (Eds.), *Educational Achievements of Thirteen-Year-Olds in twelve Countries* (pp. 63–68). Hamburg: UNESCO Institute for Education.

White, K. R. (1982). The relation between socioeconomic status and academic achievement. *American Psychological Association, 91*(3), 461–481.

Wiley, D. E., & Wolfe, R. G. (1992). Major survey design issues for the IEA Third International Mathematics and Science Study. *Prospects, 22*(3), 297–304. doi: 10.1007/BF02195952.

Woessmann, L. (2004). How equal are educational opportunities? Family background and student achievement in Europe and the US (CESifo Working Paper Series No. 1162).

Index

ACER. *See* Australian Council for Educational Research
achievement
  academic, 11, 156, 208
  educational, 8, 10, 12, 16, 22, 81, 166
  mathematics, 74, 207, 236, 248
  student, 16–17, 92, 103, 127, 165, 167
    effect of schooling on, 156
    FIMS, 10
    measure of schooling, 3–4, 7
    Model of School Learning, 106
    SES related to, 124, 162, 164, 169
    SIMS, 103, 169
    TIMSS-95, 78–79
    tracking, 206
Analyses
  benchmark, 139
  cohort, 54, 244
  cohort-longitudinal, 31, 54
  comparative, 124, 163
  curriculum, 24
  quantitative, 124
  regression, 109, 184, *See* analysis, regression
  regression, 241
  relational, 35, 51, 188, 246
  school related, 244
  secondary, 13, 36
  SIMS, 169
  statistical, 79, 188
  subtest score, 85
  textbook, 21, 130
  TIMSS-95, 144
analysis
  curriculum, 24, 28, 71, 84, 93, 95, 99, 105, 121–122, 130, 185, 214, 235, 237, 250
  regression, 143
  Test-Curriculum Matching, 98
Anderson, C. Arnold, 7–8, 167–168
applied mathematics. *See* mathematics, applied

Assessment
  curriculum-based, 216, 220–221, 222, 243, 246
  domain-sensitive, 86, 89, 91
  IEA, 4, 20, 36, 39
  international
    large-scale, 15, 164
  international comparative, 7, 20, 87, 98–99, 210
  international education, 4, 16, 36, 45
  international educational, 156, 183, 186, 210, 234, 239
  K–12, 66
  literacy, 217, 219, 245–246, 252
    mathematical, 196, 216
    problem solving, 190
  Literacy
    PISA, 116
  literacy-based, 190, 217, 220
  mathematics literacy, 242
  large-scale, 17
    international, 13–14
  mathematics, 11, 15, 49, 88, 102, 216
  performance, 31, 228
  PISA literacy, 218, 221, 230, 242
  PISA mathematical literacy, 221
  PISA mathematics literacy, 216
  quantitative, 7
  regularized, 36
  science, 67, 87, 187
  single-time point, 13
  student, 63, 92–93, 98, 104, 109, 115, 154
Australia, 63, 89, 194, 223
Australian Council for Educational Research, 29–30, 77, 308

Belgium, 63
Belgium (Flemish), 138, 175, 207
Benchmark, 130, 139
  A+, 138–139, 142, 250
  international, 138–139, 142, 250

Index

Bloom, Benjamin, 7, 105, 107–108, 168, 216
Burstein, Leigh
  SIMS, 14–16, 22, 49, 114, 188
  TIMSS, 23–24, 84, 114

Canada, 63
  PISA, 227
Carnoy, Martin, 49, 156, 166, 217, 231, 233–234, 242
Carroll, John, 106, 108, 114, 127, 216
China
  Shanghai, 184
Classroom instruction, 96, 99, 103
  implemented curriculum, 104, 142
  PISA, 119, 217, 252
  textbooks, 130, 132, 148
  TIMSS-95, 31, 35, 55, 190
Cogan, Leland, 239
  Conceptual Model of Educational Opportunities, 28
  content coverage, 214, 216, 219
  OTL Measurement, 116, 138, 142
  PISA, 54
  student background, 170
  student performance, 228
  teacher preparation, 228
  Textbooks, 130
  TIMSS, 21, 24–25, 51, 84, 91, 122, 127, 132, 154, 167, 235, 252
  total test scores, 97
  tracking, 198–199, 204, 206
  Tracking, 56
  TRENDS, 36
cognitive Olympics, 15–16, 22, 86, 101, 156
Coherence, 139, 143, 185, 235, 237
Coleman, James, xv, 65, 87, 127, 161–162, 165, 169, 210
  United States
    study of schools, 65
Common Core State Standards, 57, 129, 139, 146, 219, 236
Content coverage, 24, 57, 75, 108, 123, 130, 133, 227, 244
  IGP, 199, 201, 204, 208
  K–12 curriculum, 122
  OTL, 49, 66, 68, 122, 206, 208, 214, 216, 218, 225, 234
  PISA, 190, 196, 214
  textbooks, 130
  tracking, 57, 176, 190, 199, 202, 207–208, 219, 223
  variation in, 201–202
Content domain, 73, 87–88, 102, 123
Countries
  top-achieving, 138–141, 184, 235

Curriculum, 21, 24–25, 99, 103, 121
  attained, 28, 250
  common, 15
  Conceptual Model of Educational Opportunities, 28
  FIMS, 102
  IEA studies, 68
  implemented, 24, 26, 49, 99, 104, 106, 109, 114–115, 121–122, 133, 142–146, 155, 185, 214, 216
  index, 105
  intended, 14, 24–25, 70, 99, 104–106, 109, 114–115, 121–122, 128–130, 146, 148–154, 185, 214, 216, 250
  K–12, 80, 122, 222
  K–12, 136
  mathematics, 33, 98, 102, 116, 236
  national, 123, 126, 139, 166, 236
  OTL, 29, 30, 66, 103, 228
  Pilot Twelve-Country Study, 101, 104
  PISA, 68, 115–116, 190
  potentially implemented, 25, 28, 114, 122, 130–132, 216
  science, 105
  SIMS, 103, 114, 121, 169
  SMSO, 24
  standards, 29, 185
  structure, 146
  TIMSS-95, 25, 29, 36, 54, 99, 114–115, 122–123, 128, 190
  tracking, 206
  TRENDS, 99, 252
  variable, 21
Curriculum model
  Conceptual Model of Educational Opportunity, 32

Domain, 97
  mathematics, 74, 76, 81, 89
  overall, 81, 89
  PISA, 40, 91
  reading, 76
  science, 76, 89
  TIMSS-95, 123
  TRENDS, 89

Economic, social, and cultural status, 160–161, 172, 220
Education
  comparative, 5–9, 45, 49, 65, 101, 103, 179, 246
  international comparative, 16, 86, 101, 156, 158
  internationl comparative, 184
Educational Testing Service, 30, 40, 76–77

England, 63, 76, 82, 175
ESCS. *See* Economic, social, and cultural status
ETS. *See* Educational Testing Service

FIMS. *See* First International Mathematics Study
Finland, 82, 184, 222
  PISA, 227
First International Mathematics Study, 10–13, 82
  assessment, 14, 102
  content, 75
  database, 14
  history, 11, 20, 50, 67, 70, 74, 76, 81, 102
  items, 14
  limitations, 14
  measures, 168
  OTL, 121
  OTL measure, 102
  OTL measurment, 109
  participants, 63
  population, 245
  populations, 109
  related to SIMS, 77
  research questions, 102
  SES, 162, 164, 168
  test, 74
First International Science Study, 10–13, 81
  history, 13, 20
  OTL, 102
FISS. *See* First International Science Study
Foshay, Arthur, 6, 8, 10, 103–105, 167
framework
  TIMSS-95
    content, 78
Framework, 17, 24, 73
  assessment, 17
  common, 29
  conceptual, 7, 21, 23, 84
  content, 24, 29, 33, 70, 73, 251
  curricular content, 24
  curriculum, 124
  Curriculum Framework for Mathematics and Science, 20, 123
  IEA studies, 70, 74
  opportunity models, 36
  PISA, 39–40, 69, 72
    mathematics, 72, 91
  SIMS, 70
  TALIS, 229
  textbook, 130–131, 138
  theoretical, 100
  TIMSS-95, 70–71, 84, 93, 95, 122–124, 126, 136, 145, 237, 250

  conceptual, 24, 169
  mathematics, 93, 124, 128
  science, 71, 126, 128
France, 24, 46, 63

Gamoran, Adam, 127, 165, 198, 209, 220
Gap, 224, 245
  income, 223
  OTL, 224
  performance, 224
  research, 9
  test score, 166
GDP, 165
Germany, 29–30, 46, 63, 80, 82, 121, 129, 160, 194, 211, 230, 234, 237
  TIMSS-95
    Case Study Component, 35
    Video Study, 31
Grade
  fourth
    TIMSS-95, 85
Grade level
  seventh
    TIMSS-R, 35
  seventh
    tracking, 56
  eighth
    achievement, 204
    achievement score, 203
    aggregate score, 231
    assessment, 33, 139, 245
    assessment comparison, 230
    classrooms, 31
    cohort gain, 245
    curriculum, 90
    Curriculum Questionnaire, 153
    curriculum topics, 153
    focal, 130
    instructional practices, 35
    mathematics, 31, 198
    mathematics performance results, 185
    mean achievement, 204
    NAEP, 84
    OTL, 233–234
    OTL effect, 233
    perforance task, 33
    performance, 134, 207, 233–234
    performance score, 233
    populations, 29
    sample, 34
    student performance, 95
    student questionnaire, 169
    textbooks, 135
    TIMSS, 251
      population, 243

TIMSS-95, 83–84, 91–92
  assessment, 95
  mathematics
    topic-specific tests, 95
    performance assessment, 34
    science, 84
  TIMSS-R, 35
  tracking, 56, 198–199, 202–203, 207
  tracking effect, 202
eleventh, 244
  PISA
    modal grade, 51
end of secondary
  population, 74
  populations, 29
  years of schooling, 32
fourth
  assessment, 33
  Curriculum Questionnaire, 153
  curriculum topics, 153
  focal grade, 130
  NAEP, 84
  performance, 134
  populations, 29
  sample, 34
  student questionnaire, 170
  textbooks, 135
  TIMSS
    population, 243
  TIMSS-95, 83, 91–92
    performance assessment, 34
  TIMSS-R, 35
  TRENDS, 88
ninth, 244
  assessment comparison, 230–231
  OTL, 233
  OTL effect, 231
  performance, 89, 233–234
  PISA
    modal grade, 51
  tracking, 208
seventh
  achievement score, 203
  mathematics, 198
  pre-test, 134
  tracking, 202–203, 206–207
sixth
  performance, 89
tenth, 244–246
  cohort gain, 245
  PISA, 251
    modal grade, 51
third
  pre-test, 134
twelfth, 244

TIMSS-95, 83
twelfth grade
  end of secondary, 32
Great Britain, 46, 51
gross domestic product. *See* GDP
Gustafsson, Jan-Eric, 8–9, 11, 13, 16–17, 20–22, 28, 30, 39, 41

Hans, Nicholas, 5–6, 101
Hong Kong, 57, 132, 184–185, 194
  A+ benchmark, 138–139
Houang, Richard, 239
  Common Core State Standards, 139
  Conceptual Model of Educational Opportunity, 25, 28
  content coverage, 214, 216
  curriculum
    mathematics, 138–139, 216, 237
  national standards, 237–238
  OTL Measurement, 116, 138, 142
  SES, 164, 177, 179, 216, 223
  student performance, 228
  teacher preparation, 228
  textbooks, 130–131
  TIMSS, 21, 24–25, 51, 71, 84, 91, 93, 122, 132, 154, 235, 252
  TIMSS-95, 122, 130, 214
    curriculum, 114
    curriculum analysis, 99
    framework, 126
    textbooks, 105
    total test scores, 97
    tracking, 204
Husén, Torsten, 8
  Cognitive Olympics, 101, 156
  Comparative education, 101
  Country comparison, 80–82, 86, 90
  Curriculum, 101
  FIMS, 102, 109
    OTL measure, 102, 121
    SES, 168
  FIMS/FISS, 11–13
  IEA, 7–9, 15, 19, 22, 41, 50, 100–101, 105
    history, 17, 19
  international assessment, 20, 41, 65, 186
  International comparative study, 6
  Longitudinal study, 246
  Model
    general input-output, 103
    OTL measure, 104
  Pilot Twelve-Country Study, 10, 12
  SIMS, 14–16, 188
  TIMSS-95, 23–24, 28
Husén,Torsten
  FIMS/FISS, 11

IEA
  assessments
    curriculum-based, 216
    formation, 8–9
  studies
    approach, 161
    teacher questionnaire, 228
    variables, 221
IGP. *See* International Grade Placement index,
      See International Grade Placement
Inequalities, 225
  OTL, 178, 223–224
  SES, 178, 223–224
International Association for the Evaluation of
    Education. *See* IEA
international educational assessment, 299
International Grade Placement, 136, 138
International Grade Placement (IGP) index, 199
IRT. *See* Item Response Theory
Italy, 54, 63
Item Response Theory, 29, 78

Japan, 63, 82, 147, 184–185
  A+ benchmark, 138–139
  FIMS, 63
  Model of curriculum, 146
  PISA, 51, 227
  SIMS, 49
  SMSO, 24
  TIMSS-95, 35, 46, 129
    Case Study Component, 35
    variance, 194, 204
    Video study, 121
    Video Study, 31

Keeves, John, 77, 101, 103–104, 167
Korea, 63, 184–185, 223
  A+ benchmark, 138–139
  TIMSS-95
    variance, 204

Literacy, 67, 165, 241
  definition, 40
  mathematical, 213, 218, 240–244, 252–253
  mathematics, 219, 234
  performance, 116, 120, 218, 233, 247
  PISA, 40, 55, 72, 234
    definition, 91
    mathematical, 68–69, 72, 91, 119,
        222–223, 226
    science, 69, 91, 213
Literacy assessment
  TIMSS-95
    mathematics, 32
    science, 32

Martin, Michael, 36, 75, 78, 88, 92
Mathematical literacy. *See* Literacy:PISA,
    mathematical
Mathematics
  formal school, 241
Mathematics Framework. *See* framework,
    mathematics
McKnight, Curtis
  Assessment, 87–88, 95–98
  Conceptual Model of Educational
      Opportunity, 28
  content coverage, 214, 216
  curriculum, 154, 176, 235
  OTL Measurement, 116, 127, 138, 142, 179
  SMSO, 24, 28, 36
  textbooks, 130
  TIMSS, 21, 24, 51, 71, 84, 91, 93, 122, 132,
      235, 252
  TIMSS-95, 21, 28, 71, 89, 115, 122–124,
      129–130, 214, 235
    curriculum, 114
    curriculum analysis, 99
    textbooks, 105
  Total test scores, 97
  tracking, 204
measure
  student achievement, 104
Measures
  international, 3, 66
Methodology
  conditioned multiple imputation, 79
  large-scale survey research, 65
Methods
  mixed, 35
  plausible value, 79
  psychometric, 159
  quantitative, 14, 142
  schooling, 5
  scientific, 6
Model
  Conceptual Model of Educational
      Opportunity, 24, 27, 28
  Estimated structural model of curriculum,
      147
  IEA Tripartite Curriculum, 21, 25, 104–105
  of Mastery Learning, 105, 107
  of Mathematical Literacy, 68
  of Potential Educational Experiences, 115
  of School Learning, 106–108, 114, 216,
      251
  of SES Inequality, 221
  Rasch, 97
Modeling
  Item Response Theory, 78, 84, 97, 252
Mullis, Ina, 36, 88, 92–93, 154

Index

NAEP. *See* National Assessment of Educational Progress
National Assessment of Educational Progress, 46, 77, 84
National Center for Educational Statistics, 83, 159
NCES. *See* National Center for Educational Statistics
Netherlands, 63, 175, 178, 194, 223
New Zealand, 51, 63
Noah, Harold, 5
Norway, 24, 34, 63, 177
   TIMSS-95
      variance, 204

OECD
   backgound, 37
   Education study history, 38
   studies
      approach, 161
Opportunity to learn
   applied mathematics, 120
   applied problems, 119, 217
   construct, 104
   data, 33, 51, 108–109, 115, 132, 247
   definition, 142
   defintion, 12
   formal school mathematics, 116, 119, 217, 243
   indicator, 102, 109, 145
   measure, 14–15
      IGP, 138
      related to
         student performance, 103
   measure of schooling, 4
   measures, 98, 104, 106, 108–109, 114, 148, 163
      quantitative, 28, 127
   measures of
      qualitative, 28
   performance expectations, 130
   PISA
      measurement, 120
   questionnaire, 145
   related to
      achievement, 28
      performance, 132
      socio-economic status, 124
      student performance, 29, 216
   relationship to
      achievement, 29, 153
      performance, 109, 124, 177, 216–217, 224, 228, 230–231
   variable, 102, 233
   variation in, 201, 214, 216, 246
   within country, 214
   word problems, 119
Organisation for Economic Co-operation and Development. *See* OECD
OTL. *See* Opportunity to learn

Performance
   country, 90, 123, 139, 144, 156
      mathematics, 78
   data
      student, 96
   mathematics, 185, 205, 224, 227, 241, 243
   student, 55, 79
      mathematics, 243
   related to
      cumulative content coverage, 133
      tracking, 198
   variance
      student, 109
Pidgeon, Douglas A., 8, 10, 103, 167
Pilot Twelve-Country Study, 9, 12–13, 20, 34, 63, 65, 67, 74, 76, 100–101, 103–106, 109, 167, 185
PISA
   assessment, 40
   focus, 39
   history, 39
   IEA
      difference, 40
   literacy assessment, 216
   OECD relationship, 36
   studies
      SES index, 220
   study of teachers, 228
Population
   PISA
      15-year-old, 243
   student, 31, 74, 79, 104, 160, 166, 170, 185, 246
Postlethwaite, T. Neville, 8
   FIMS/FISS, 11–13
   IEA, 7–9, 15, 19, 22, 41
      history, 17, 19
   international assessment, 20
   SIMS, 14–16
   TIMSS-95, 23–24, 28
   Twelve-Country Pilot Study, 12
Practices
   classroom, 49, 228
   educational, 54, 246
   instructional, 4, 21, 31, 67, 183, 188, 228, 248
   leadership, 229
   teaching, 228, 230

Processes
 classroom, 49, 80, 114, 188
 instructional, 31
 mathematical, 72
Programme for International Student Assessment. See PISA

Raizen, Senta, 23
 Conceptual Model of Educational Opportunity, 28
 curriculum analysis, 99
 TIMSS-95, 21, 24, 71, 89, 115, 122–124, 129–130, 214, 235
  curriculum, 114
  textbooks, 105
Robitaille, David, 23
 SIMS, 82, 109, 168
 TIMSS-95, 21, 24, 71, 123–124
Russia, 51, 211, 230, 242, 245, 249
 longitudinal study, 217, 230–234

Sampling
 design, 54–55, 246
  early IEA studies, 108
  matrix, 74, 78, 91
  PISA, 115, 196, 215, 247
  TIMSS, 247
  TIMSS-95, 32, 134, 197
  tracking, 209, 219–220
  two-stage PPS random, 144
  variance components, 196
  variation, 193
 errors, 11, 93
 frame, 51, 57, 220, 246–247
 procedures, 92, 246
 tracked system, 55–57
 unit, 144
  IEA studies, 55, 246
  PISA, 55, 145
  TIMSS-95, 145
  TIMSS-R, 36
Scaling
 Item Response Theory, 29, 79, 88, 252
Schmidt, William, 23, 239
 Assessment, 87–88, 95, 97–98
 causal inferences, 49, 217
 Common Core State Standards, 139
 Conceptual Model of Educational Opportunity, 25, 28
 content coverage, 214, 216, 219
 curriculum, 24, 28, 114–115, 142, 154, 167, 176, 235
  mathematics, 138–139, 216, 237

 mathematics education, 241
 national standards, 237–238
 OTL Measurement, 116, 127, 138, 142, 179
 PISA, 54
 Russian longitudinal study, 231, 233, 242
 SES, 164, 177, 179, 216, 223, 226–227
 SMSO, 24, 28, 36
 student background, 170
 student performance, 228
 teacher preparation, 228
 textbooks, 130–131
 TIMSS, 24–25, 51, 71, 84, 91, 93, 122, 127, 132, 154, 167, 235, 252
 TIMSS-95, 21, 24, 28–29, 71, 89, 115, 122–124, 126, 129–130, 138, 142, 145–146, 179, 214, 235, 248
  curricululm, 114–115
  curriculum analysis, 99
  framework, 126
  textbooks, 105
  topic trace mapping, 154
 Total test scores, 97
 tracking, 56–57, 198–199, 204, 206
 TRENDS, 36
Schooling
 K–12, 4, 243–244
 measure of, 123, 126, 142, 197, 251
 role of, 3, 154, 157, 210–211, 230, 245
 study of, 68, 156, 246–247
Schwille, John, 11, 15, 22–23, 156
science framework. See Framework, science
score
 percent correct, 88–89, 91–93
 Percent correct, 92
 total scaled, 65, 87–88, 95, 99
 total test, 81–82, 95, 132, 252–253
  comparisons, 84
  subdomain, 91
Second International Mathematics and Science Studies. See SIMS, See SISS
SES.
 composite index, 161–162
 definition, 158–159
 measure of
  educational background, 160
  IEA studies, 166–170
  schooling, 4, 156
  occupation, 159–160
  PISA studies, 170–172
  wealth, 160–161
 related to
  OTL, 33–35, 172–177

performance, 33–35, 124, 164, 175, 177, 230–231
role
　international assessments, 162–166
SIMS, 13–15
Singapore, 47, 82, 88, 184–185
　A+ benchmark, 138–139
SISS, 13–15
Six Subject Survey
　databases, 14
　history, 13, 20, 67, 105–106
Slovenia, 147
　Model of curriculum, 146
socio-economic status. See SES
Spain, 24, 54
Standards
　curriculum, 96, 102, 105, 225
　K–12 curriculum, 93
Student Achievement
　PISA, 232
　TRENDS, 232
Studies
　international comparative, 156, 158
　Performance Assessment, 33–35
Subdomain
　IEA studies, 90
　scores, 82
Subdomains, 81
　PISA, 88
　SIMS, 82
　TIMSS-95, 84
　TRENDS, 88
Sweden
　TIMSS-95
　　variance, 204
Switzerland, 24

Taiwan, 185
TALIS. See • Teaching and Learning International Survey
Teacher Education and Development Study in Mathematics, 66, 227–228
Teaching and Learning International Survey, 66, 228–229, 247
TEDS-M. See Teacher Education and Development Study in Mathematics
Textbooks
　coding, 131, 136
　emphasis, 132, 134–135
　potentially implemented curriculum, 25–26, 122, 130, 214
Third International Mathematics and Science Study. See TIMSS-95

Thorndike, Robert L., 8, 9–10, 66, 103, 105, 167–168
TIMSS-95
　Case Studies component, 35
　defined, 21
　OTL measurement, 29
　OTL relationship to performance, 29
　populations, 23
　Secondary School Study
　　Generalists, 32
　　Specialists, 33
　Video Study, 31
TIMSS-Repeat. See TIMSS-R
Topic trace mapping
　mathematics, 136
Topics
　mathematics, 95, 102, 129, 140, 153, 241, 250
　science, 141, 250

UNESCO, 7, 160
　Institute for Education, 8–9, 65
United Nations Educational, Scientific and Cultural Organization. See UNESCO
United States, 80, 82, 147, 211, 230, 234, 237
　content coverage, 129
　data, 145, 197–198
　　TIMSS-95, 57
　FIMS, 63
　government, 23, 45–46, 77, 83–84
　IEA
　　data processing center, 9
　intended curriculum, 129
　Model of curriculum, 146
　national standards, 128, 139
　OTL
　　tracking, 202
　PISA, 51, 54
　scores, 166
　SES inequality
　　tracking, 177
　standards, 129
　students, 208, 235
　textbooks, 131, 148
　TIMSS-95, 84, 194
　　Case Study Component, 35
　　reports, 29
　　sample, 197
　　Secondary School Study, 33
　　Video Study, 31
　tracking, 56, 190, 196, 199, 219
　　sampling, 57
　TRENDS, 88, 175
　variation, 214

US
  TIMSS-95, 46, 129
    Video study, 121

Variance component, 191, 193, 197, 201, 204, 246
  classroom, 194, 197
  corresponding, 193
  estimates, 196
  schools, 194
  tracking, 57, 197
  within-classrooms, 196
  within-school, 196

Wall, W. D., 8
Wiley, David, 23
  Conceptual Model of Educational Opportunities, 28
  content coverage, 214, 216
  OTL Measurement, 116
  textbooks, 130

TIMSS, 21, 24, 51, 71, 84, 93, 122, 127, 132
TIMSS-95, 122, 130, 214
  curriculum, 114
  curriculum analysis, 99
  design, 29, 49, 189
  textbooks, 105
  total test scores, 97
  tracking, 56, 198–199
Wolf, Richard, 8
Wolfe, Richard, 23
  Conceptual Model of Educational Opportunity, 25, 28
  content coverage, 214, 216
  curriculum analysis, 99
  textbooks, 130
  TIMSS, 21, 25, 51, 84, 122, 132
  TIMSS-95, 24, 71, 122, 130, 214
    curriculum, 114
    design, 29, 49, 189
    textbooks, 105
    total test scores, 97